COMMUNICATION SKILLS AND PERSONALITY DEVELOPMENT

ABOUT THE AUTHOR

Dr. J. R. Kadam, Associate professor of Extension Education, College of Agriculture, Dapoli has participated in over15 National, State level training programs, and 10 National level seminars. He has delivered two T.V. and four Radio talks. He has mentored nine M.Sc. and a Ph.D. (Agri. Extn.) scholars and was also was invited as a resource person at *'21 days winter school'*. Dr. Kadam has over 17 research articles, 5 research notes, 20 research abstract and 4 books to his credit.

Dr. V. G. Patil, Professor and Head of Extension Education, College of Agriculture, Dapoli has worked on various positions in the University. In his distinguished career he has guided 35 M.Sc. students and 5 Ph.D. scholars. He has participated in several national and international seminars and presented 30 technical papers in his capacity. Dr. Patil has also developed 25 training manuals for farmers and extension functionaries.

Mr. S. A. Dhenge, Ph.D. scholar, Department of Extension Education, College of Agriculture, Dapoli. He has qualified ICAR NET exam twice. He has one book chapter, a book, 15 research abstracts, 14 research papers and over 51 popular articles to his credit. Mr. Dhenge has participated in several International and National seminars, and National level training programs; he has been awarded with the *'Best Author Award'* from *'Krishidoot magazine'*. He has delivered one Radio talk.

Mr. A. M. Murai, Ph.D. scholar, Department of Extension Education, College of Agriculture, Dapoli. He has published 5 research abstracts, 11 research papers and 2 popular articles in the field of Agriculture. He has participated in several National and International Seminars, and National level training program.

COMMUNICATION SKILLS AND PERSONALITY DEVELOPMENT

J.R. Kadam
V.G. Patil
S.A. Dhenge
A.M. Murai

SCIENTIFIC PUBLISHERS

Published by:

SCIENTIFIC PUBLISHERS (INDIA)
5A, New Pali Road, P.O. Box 91
Jodhpur 342 001, India
E-mail: info@scientificpub.com
Website: www.scientificpub.com

Print: 2024

ISBN: 978-93-87741-68-7
eISBN: 978-93-87991-76-7

Printed in India

FOREWORD

A systematic human resource management approach based on available tools and techniques is the need of the hour in public extension organizations. The public extension system must institutionalize the best of human resource practices and functions of training and development, career planning, motivation, leadership development and performance. Therefore plenty of lessons to be learnt from the other public service organizations. Considering the right future of extension service, personality is an important thing which determines not only professional success but also determine overall behavior and attitude of individual. The person having a good personality and better communication skills can move through the difficulties with more confidence. In this context, this book provide relevant and essential information regarding the communication skills and personality development.

I am sure that the book will help all those engaged in education, agricultural extension, mass communication and similar others fields in making their job more effective. I congratulate the authors for their valuable work.

Dr. U. V. Mahadkar

Associate Dean

College of Agriculture, Dapoli
Dr. B. S. Konkan Krishi Vidyapeeth
Dapoli, Ratnagiri (M.S.)

PREFACE

Personality development has become an important tool today for developing overall skills within a person that help him to develop professionally as well as personally. A great personality includes knowing how to dress well, social graces, grooming, speech and interpersonal skills. There are many factors like biological characteristics, family and social groups, cultural and social factor, which contributes towards formation of personality. On the other hand, good communication is vital to any institution's successful operation and equally imperative for a man's personality development. This book is a thorough attempt to present these concepts in simple, student-friendly language so that difficult situations can be handled in an easy way.

The course 'Communication skills and Personality development.' has been recommended by V Deans committee for B.Sc. (Agri.), B.Sc. (Horti.) and B.Tech. faculty in all agricultural universities in India. The book has been directed to cover the entire syllabus of this course, "so that the book can be used as a textbook for the under graduate Agricultural students".

We sincerely acknowledge with gratitude the inspiration and encouragement we have received from Dr. U. V. Mahadkar, Associate Dean, College of Agriculture, Dapoli, Dist. Ratnagiri (M.S.). We are also grateful to all those who have helped us directly and indirectly in bringing this publication in its present form.

Authors

LIST OF ABBREVIATIONS

TAT	Thematic Appreciation Test
KISS	Keep It Short And Simple
e.g.	Example
HR	Human Resources
CP	Controlling Parent
NP	Nurturing Parent
AC	Adaptive Child
NC	Nurture Child
TA	Transactional Analysis
LP	Little Professor
POSEC	Privatize by Organizing Streamlining Economizing and Contributing
BTW	By The Way
LOL	Laugh Out Loud
ELC	Experiential Learning Cycle
HRD	Human Resource Development
TWT	Training Within Industry
TM	Transcendental Meditation
Ed.	Education

CONTENTS

CHAPTER 1
PERSONALITY

CONCEPT OF PERSONALITY

Personality is the total quality of an individual. The word personality comes from the Latin '*persona*' which means the mask worn by players in the theater. Personality consists of observable behaviour. It is defined as an individual, typical or consistent adjustments to his environment. The elements of personality are called the traits of personality, it is the traits that make one person different from another person in their behaviour. Shyness and sociability are different traits of personality. Personality is the dynamic organization of the traits and characteristic of an individual. All characteristics which an individual possesses are his powers, needs, abilities, wants, habits, his goal and aspirations. His patterns of behaviour to objects and persons shape his personality.

Since an individual is a bundle of characteristics and traits we can define personality as an integrated pattern of traits.

DEFINITION OF PERSONALITY

- Personality is the sum total of ways in which an individual interacts with people and reacts to situations.
- Personality is that which makes one effective, or gives one influence over others. In the language of psychology it is one's social stimulus value —**May, M. A.**
- A man's personality is the total picture of his organized behaviour, especially as it can be characterized basis fellow men in a consistent way —**Dashiell, J. F.**
- Our personality is the result of what we start with and what we have lived through. It is the 'reaction mass' as a whole —**Watson, J. B.**

- Personality is the dynamic organization within the individual of those psychological systems that determine his unique adjustments to his environment —**Allport, G. W.**

It can, therefore be defined as an individual's typical or consistent adjustments to his environment. It is the most characteristic integration of an individual's structure, modes of behaviour, interests, attitudes, capacities and aptitudes.

PERSONALITY TRAITS

- **Extraversion:** Refers to the extent to which a person is comfortable with other people. People who have a high degree of extraversion are sociable, talkative, and friendly. They tend to prefer jobs that require them to interact with a number of different people - HR, marketing, PR, etc. People who are low in extraversion are called introverts. They prefer to be by themselves instead of talking to others. They are reluctant to interact with other people and avoid developing new relationships.
- **Agreeableness:** Refers to the extent to which a person subjugates his interests for the sake of the group. People who are very agreeable give importance to maintaining harmony and do not insist that others agree with what they say or follow their suggestions. Agreeable people are good-natured, cooperative and trust others. They are likely to develop good working relationships with all organizational constituents (coworkers, subordinates, superiors, customers, and suppliers). They can be used as harmonizers of interest groups and self-help groups. People who are less agreeable give more importance to their own needs, opinions and values than those of others. They are unlikely to develop good relationships.
- **Conscientiousness:** Refers to the extent to which a person is responsible and achievement oriented. People who are very conscientious limit the number of goals they set for themselves, devote their time and energy to those goals, and often succeed in achieving their goals. They are responsible, dependable, persistent and highly achievement-oriented. They are more organized, responsible and self-disciplined and perform better

in their jobs than those who are not so conscientious. People who are less conscientious, set too many goals for themselves and often fail to achieve any of them.

- **Emotional stability:** Refers to the individual's ability to withstand stress. Individuals who have positive emotional stability feel emotionally secure and tend to be calm. They are enthusiastic about their work and are capable of withstanding the tensions and pressures of a job. Such farmers should be selected as model farmers. People who have negative emotional stability are emotionally insecure and experience feelings of anxiety, nervousness, and depression. They are unable to withstand job pressures.
- **Openness to experience:** Refers to an individual's range of interests and indicates how innovative or how rigid he is in his beliefs. An individual with high level of openness tends to be creative and had a wide range of interests. Such individuals are open to learning and make good workers. Individuals who have low level of openness have a narrow range of interests, rigid mindsets and tend to be less curious and willing to accept new ideas. Such individuals make poor workers.
- **The Self-Concept (self-esteem and self-efficacy):** 'Self' refers to the personality of an individual as viewed by that person himself. Self-concept refers to the efforts made by an individual to understand his own self. Self-concept is closely related to the concepts of self-esteem and self-efficacy.

  a. ***Self-esteem:*** Self-esteem refers to the self-perceived competence and self-image of people. It has a moderating influence on employees' emotional and behavioural responses to various situations and the stress experienced by them. Employees with high self-esteem perceive themselves as unique, competent, secure and empowered. They possess the ability to positively influence situational factors in order to accomplish the assigned tasks. Such people are able to confidently and freely interact with people around them. Self-esteem is a generalized trait (it is present in all situations).

b. ***Self-efficacy:*** Self-efficacy refers to a person's perception of his ability to cope with different situations as they arise. People with high self-efficacy have the capability and the required confidence to rise to the occasion. Self-efficacy is situation specific. Self-efficacy and employee performance are highly correlated. The relationship between self-efficacy and performance is cyclical — self-efficacy affects performance and performance affects self efficacy. The role of self efficacy is vital in helping an individual cope with tough jobs, make a career choice, learn and achieve something and adapt to new technology. Organizations can enhance self efficacy of employees by training them. The self-efficacy of a person can be measured along three dimensions — level, strength and generality. Level refers to the number of tasks a person can effectively perform. Strength refers to how firmly an individual believes he is capable of performing a task.

PERSONALITY TYPES

According to Jung

- **Introverts** will react negatively to situations, withdraw from the society, inwardly, selective. The introvert withdraw from the active participation in the objective world and he is interested in his inner world of thought and fantasy. He will not move freely with others. He won't express out his feelings and inner desires. He is sensitive to criticism, magnifies his failures and occupied with self-criticism.
- **Extroverts** are opposite to introverts. Extroverts react positively and outwardly expressive. An extrovert is supposed to be thick skinned and relatively sensitive to criticism, spontaneous in his emotional expression, impersonal in argument, neither deeply affected by failures nor much occupied with self-analysis or self-criticism.

Sheldon's Classification

Sheldon classifies people into three types.

1. Endomorphy 2. Mesomorphy 3. Ectomorphy.

In addition to this, Sheldon also classified people into three types according to their temperaments. They are (1) Viscerotonia, (2) Somatotonia and (3) Cerebrotonia as discused in Table 1.

Table 1: *Sheldon's classification according to temperaments.*

Sr.No	Body types	Temperamental types
1.	**Endomorphy** Large viscera Soft body contours	**Viscerotonia** Love of comport affection |Sociability
2.	**Mesomorphy** Heavy muscular development Amlitious Hard body contours	**Somatotonia** Vigorous, self- assertive
3.	**Ectomorphy** Long, slender inhilated Extremities, poor muscular development	**Cerebrotonia** Restrained Social withdrawal thoughtful

Measurement of Personality

The following methods have been used for measuring personality characteristics.

- **The Questionnaire Method:** This method involves the preparation of a list of questions or statements. Each one is concerned with some aspect of the feeling, attitude, habit or mode of behaviour related to the personality characteristic which is intended to be measured. The respondent is required to indicate his agreement or disagreement, acceptance or rejection, affirmation or denial of each statement. The questions are constructed in such a way that the affirmation of some and the denial of others express the presence of the trait that is measured. Each question is followed by 'Yes', 'No', 'Untrue', 'Doubtful' or other similar responses. Here the respondent under line a response that shows his character.

  Example:

  a. In social gatherings, I like to be the centre of attention.

  b. In a meeting, I prefer to occupy a back seat.

  c. When you have to make an important decision, do you prefer someone else to decide for you?

d. Usually a personality test contains 30 to 50 items.

e. This questionnaire method is also called Objective Test of personality. In this method the respondent should be truthful and very frank.

- **Projective Tests:** Projective tests are so named because they induce the individual to project, to put himself into the test situation or to identify with the person therein and by telling about them, to reveal his own motives, attitudes, apprehensions and aspirations. The use of projective test is based on the mechanism of projection. The projective test is a device for measuring the personality which the person tested on, does not recognize as such. He makes spontaneous responses in the test. The test presents a highly ambiguous situation which the rephrase is required to perceive and describe. The situation has no definite characteristics and so different persons perceive it in different ways. In perceiving the ambiguous situation, each person reflects his own needs, attitudes, habits, interests, feelings and behaviour patterns. The Projective method is called projective because in perceiving and describing the situation, a person brings out his personality characteristics. Two well-known projective tests are (a) Roschach Inkblot Test and (b) Thematic apperception Test.

  a. **Rorschach Inkblot Test:** The test was first described in 1921 by Hermann Rorschach, a Swiss psychiatrist. This is the most widely used test. Rorchach Test consists of 10 cards, each one having an inkblot on it. Some blots are coloured and some are in black and white. These cards are always presented in regular serial order. The inkblot does not represent any subject. It is ambiguous or unstructured figure. The ambiguity of the blot results in the great variety of responses it produces. Such an inkblot is placed before the subject. He is asked to describe what he sees. He is also required to say in what portion he sees and what makes him see the object that he describes. His responses are noted and classified under different categories. These responses differ from person to person. Some respond to the whole figure, some to the parts of the figure. Some respond to form a shape, some to colour or the white space between the coloured or black patches. Some see moving and others stationery objectives. Some see human figures or parts of human body. Others see animals. Classification of the responses made to the 10 cards by a subject shows that certain types and categories are made consistently. These responses are interpreted as indicating the personality characteristics of the respondent.

b. **Thematic Apperception Test (TAT):** This TAT is also widely used projective Test. TAT was first given by Murray. This test consists of 20 pictures. Each picture contains one or more persons in very ambiguous situations. For example, in one picture a very old woman is standing behind a young woman looking very serious. The subject is asked to write a story on each picture. The study should have a theme. It should say what is happening, and what is going to be the outcome or result. In writing a story, the subject is expected to identify himself unconscientiously with a character in the story. The story thus express out his own needs and frustrations, feelings and attitudes, ideas of self and of others, real or imaginary and so on. One cannot get clue about the characteristics of the subject from a single story. When one finds the recurrence of the same or similar characters, expressing similar feelings and attitudes, entertaining similar hopes and fears etc., one may be led to thinking that the stories reveal the person's own characteristics.

- **Personality Rating:** In this method one a person judges or rates the characteristic of another person. The person who judges is the "rater and the person who is judged is the rate". The basis of the rating is the rater's general impression drawn from his observation of the behaviour of the rates in a variety of related situations. For example, a class teacher may be asked to judge such personality characteristics of his pupils as order lines, punctuality, industriousness, co-cooperativeness etc. The teacher may be asked to express his rating on each trait by locating the position of each pupil on a so called rating scale.

- **Situational Test:** In the situational test, a person is required to act in a situation which is specially arranged for the purpose of testing. The tester observes the behaviour of the tested while he is acting in the situation. Generally, the testing is done in a group. For example, a party of 10 students is taken out and camps at the foot of hill. The students are given some cash. They are asked to plan and arrange for preparing their lunch. In this they will discuss and plan to report lunch. The teacher who accompanies them observes the behaviour of each and come to the conclusion regarding their personality character.
- **Interview:** Interview is one of the methods to assess personality. The interview is a face to face situation consisting of the interviewee and the interviewer. The success of the interview depends upon the degree to which the interviewee makes free and frank responses. The interviewer should win the complete confidence of the interviewee. He should establish with the interviewee a relation of cordiality, warmth and responsiveness. Such a relation has been called 'rapport' without a rapport no psychological interview can be a success.

ANALYZING STRENGTH AND WEAKNESS

Origin and Development

The development of the strengths and weaknesses analysis can be traced back to a planning trend that presented itself in American business and industry in the mid twentieth century.

It was probably initiated by Dupont that started making long-term strategic plans in 1949. You are most likely to succeed in life if you use your talents to their fullest extent. Similarly, you'll suffer fewer problems if you know what your weaknesses are, and if you manage these weaknesses so that they don't matter in the work you do. So how you go about identifying these strengths and weaknesses, and analyzing the opportunities and threats that flow from them? SWOT Analysis is a useful technique that helps you do this.

Strengths

- What advantages do you have that others don't have (for example, skills, certifications, education, or connections)?

- What do you do better than anyone else?
- What personal resources can you access?
- What do other people (and your boss, in particular) see as your strengths?
- Which of your achievements are you most proud of?
- What values do you believe in that others fail to exhibit?
- Are you part of a network that no one else is involved in? If so, what connections do you have with influential people?

Weaknesses

- What tasks do you usually avoid because you don't feel confident doing them?
- What will the people around you see as your weaknesses?
- Are you completely confident in your education and skills training? If not, where are you weakest?
- What are your negative work habits (for example, are you often late, are you disorganized, do you have a short temper, or are you poor at handling stress)?
- Do you have personality traits that hold you back in your field? For instance, if you have to conduct meetings on a regular basis, a fear of public speaking would be a major weakness.

Opportunities

- What new technology can help you? Or can you get help from others or from people via the Internet?
- Is your industry growing? If so, how can you take advantage of the current market?
- Do you have a network of strategic contacts to help you, or offer good advice?
- What trends (management or otherwise) do you see in your company, and how can you take advantage of them?
- Are any of your competitors failing to do something important? If so, can you take advantage of their mistakes?
- Is there a need in your company or industry that no one is filling?

Threats

- What obstacles do you currently face at work?
- Are any of your colleagues competing with you for projects or roles?
- Is your job (or the demand for the things you do) changing?
- Does changing technology threatening your position?
- Could any of your weaknesses lead to threats?

CHARACTERISTIC OF PERSONALITY

- **Psychophysical Systems:** Personality is a system that has both psychological and physical characteristics. This system is composed of interacting elements and the main elements of the system are traits, emotions, intellect, temperament, character and motives. All these elements are psychological, but they are based in the neurology and endocrinology of the body.
- **Dynamic Organization:** It signifies that different elements of psychological system are independent but function in an interlocking manner and are subject to change. Though this change can take place over a period of time in a gradual manner. The above examples may seem to you quite radical, but changes in the personality of all individuals do come as they enter new roles, responsibilities and circumstances.
- **Consistency:** Since personality is a stable organisation it also has the element of consistency. Through consistency we mean that an individual behaves in the same method in different situations and behavioral consistency is found when same situation is repeated crossways time. Psychologists provide four types of consistency.
    a. ***Type–A Consistency:*** In this type the situation and behaviour remain same. For instance, a football player is asked to play match against exam X and after 10 days he asked to play against the team.
    b. ***Type–B Consistency:*** When the same behaviour is repeated in two different situations. For instance, a person appears for an interview before one selection board then he appears

for interview for the same job before another board in the final round.

c. ***Type–C Consistency:*** When an individual is asked to behave differently in the same situation. For instance, when an actor is asked to imitate different actors.

d. ***Type–D Consistency*:** When a person behaves differently in different situations.

- **Unique Adjustment to Environment:** Every person is characterized with a dynamic organisation of psychological traits that creates his adjustment. The cause for this is that experiences of every person are unique so their reaction to the environment is also unique. You may notice that even identical twins who come out of the same embryo, though have the same genetic make-up, react differently to the same situation because their frame of references is unique.
- **Development of Personality Structure:** Personality development is the natural quality of a rising organism. The path is from simple to increasingly intricate factors and situations an individual has to pass through. According to Heniz Werner, at birth the mental organisation of the infant expands slowly. Through interactions with the environment, the parts of the child's mental structure become progressively crystallized and differentiated from each other. The analytical stage is followed through synthesis or integration when the differentiated parts become functionally organized. From a diffused mass through progressive differentiation to an integrated whole is, then the course of development of personality structure.
- **Consciousness:** Personality is conscious in that it develops out of our interaction with the environment. This interaction results in formation of concept of self. Self-concept means who we are and what we stand for. All the responses of a human being are oriented toward protection of the self-concept.
- **Potentiality for Change:** Potential for change is another characteristic of personality. The earlier psychoanalytical view did hold personality as a rigid structure. Though, modern

humanistic theories have demonstrated not only the human has the capability for reorganization but also the circumstances do foster change. Integration or organisation is the quality of the human personality that is it occurs to human beings naturally and normally. It is the normal development outcome of personality structure. Disorganization, that is, the isolation of the functions of the individual parts from the total system, is a pathological condition (Goldstein) of a psychological disorder.

Personality Determinants

- **Heredity (Nature):** The biological, physiological or psychological characteristics that an individual is born with constitute heredity. Characteristics that an individual may partially or wholly inherit from either of the parents are physical stature, facial features, skin and hair colour, temperament, muscle composition and reflexes, energy levels and biological rhythms.
- **Environment (Nurture):** Environment refers to the culture of the society in which an individual is brought up; the norms set by parents, teachers, significant others and social groups with whom the individual interacts; and other situations and experiences that the individual undergoes in his life.
- **Situation (it depends):** Different situations bring out different aspects of an individual's personality. Example, a person will attempt to control his behaviour when interacting with the MD of the company; however, when he is among his friends he will be relaxed and make minimum efforts to control his behaviour.
- **Machiavellianism:** Refers to the extent to which an individual is practical in his approach, maintains an emotional distance from others, and believes that the ends justify the means. Those who score high on Machiavellianism are good at manipulating others and try to win by any means. They do not need to be persuaded to work but instead are able to successfully persuade others. They perform well in situations that involve face-to-face meetings. They are especially productive in jobs

that require the use of bargaining (persuasion) skills and in jobs that offer substantial rewards for the achievement of goals.

- **Self-esteem:** Refers to the degree of liking an individual has for himself. People with high self-esteem are generally confident that they possess the abilities required for succeeding at work. They prefer to take up unconventional or challenging jobs. They do not care about pleasing others and fail to be influenced by external factors. They are not afraid of taking unpopular stands.
- **Self-monitoring:** Refers to the ability of an individual to adapt his behaviour to the demands of the situation. High self-monitors are capable of changing their behaviour to the demands of the situation. They can play multiple and even contradictory roles. The way they behave with their employees, boss, clients, and suppliers is entirely different in each case. They make successful managers and tend to get promoted faster than others.
- **Risk-taking:** Refers to the extent to which an individual is prepared to take risks. Such behavior is required to develop entrepreneurship in agriculture high risk-takers make decisions very quickly without searching for much information. Risk-averse people do not make decisions in a hurry and gather a lot of information before making any decision.

Theories of Personality Development

- **Levinson's Theory of Adult Life stages:** Daniel Levinson proposed that the personality development of an individual progresses with age. He theorized that there are four transitional periods and four periods of stability. Personality development takes place during the transitional periods; during the periods of stability, no development takes place. The maximum development of one's personality takes place during the mid-life transition. An individual's attitude towards work undergoes tremendous change during periods of transition rather than during periods of stability. There is, however, no evidence to back up Levinson's theory. Since he was not able to explain the vast differences in personality among people of the same age, he redefined the stages as eras — early adult, mid-life, and late adult — each of which included a transition-in period, a period of stability, and a transition-out period. He later modified his approach to suggest that every stage of an individual's life cycle is characterized by interplay of mobility and stability.

Table 2. *The transitional periods and periods of stability are as under.*

Periods of stability Periods of transition	Periods of stability Periods of transition
Stepping into the adult world: Age22-28	Age 30 transition: Age 28-33
Settling down: Age 33-40	Mid-life transition: Age 40-45
Stepping into middle adulthood: Age 45-50	Age fifty transition: Age 50-55
Culmination of middle adulthood: Age 55-60	Late adult transition: Age 60-65

- **Hall's Career Stage model:** Hall blended Levinson's theory with other adult stage theories to develop a model for career stages.

Stage	Activity
Exploration stage	• Seeks an identity for himself • Attempts to understand himself and his personality • Tries out various roles in his career • Relatively unstable in his career and less Productive • Keeps switching jobs in search of the right one
Establishment stage	• Employee tries to settle down in his job • Interacts with co-workers to develop a good relationship with them • Productivity increases
Maintenance stage	• Productivity reaches its peak • Feels the need to contribute something to the next generation (may act as a mentor to his Subordinates) • Productivity may either increase or remain stagnant
Decline stage	• Productivity starts declining • Individual evaluates his life and tries to convince himself that he made the right choices

- **Chris Argyris' Immaturity to Maturity Continuum:** Proposed that the degree of development of an individual's personality can be explained in seven dimensions-
  a. Passivity activity
  b. Dependence- independence
  c. Behaving in a few ways - Capable of behaving in many ways
  d. Erratic, shallow interests- Deeper and stronger interests
  e. Short time perspective- Long time perspective (past and future)
  f. Subordinate position - Equal or superordinate position
  g. Lack of awareness of self - Self-awareness and control

  Awareness and control over self Just as a human being develops 'from an infant to an adult, human personality development too takes place along a continuum, rather than stages, from immaturity to maturity. It is not essential that all individuals exhibit all the seven dimension of personality on reaching the mature end of the continuum.

Table 3. *Immature and Mature characteristics.*

Immaturity characteristics	Maturity characteristics
Passivity	Activity
Dependence	Independence
Few ways of behaving	Diverse ways of behaving
Shallow interests	Deep interests
Short-time perspective	Long time perspective
Subordinate position	Superordinate position
Lack of self awareness	Self-awareness and control

The seven dimensions reflect only one aspect of an individual's personality. Personality also depends on other factors like perception, self-concept, and the ability to adapt or adjust. As an individual progresses from infancy to adulthood along the continuum, there is a continual change in the level of development along different dimensions This model can only measure and describe the development of an individual's personality but cannot predict any specific behaviour of the individual. The latent characteristics of personality, which form the basis of the seven dimensions, may be quite different from the externally visible behaviour of people. The personalities of employees in an organization are generally at the mature end of the continuum. However, organizations fail to recognize this maturity and treat employees as if they are immature. This leads to conflict. Therefore farmers with high level of maturity should be used as contact or model Farmers.

- **Edgar Schein's Socialization Process:** Socialization refers to the process through which an individual's personality is influenced by his interaction with certain persons, groups, and society at large. It is a continuous process that goes on life-long. Since socialization has a major impact on the behaviour of employees in organizations, Schein proposes that it is important to analyze and control the socialization forces in an organization. The socialization process in an organization is mostly confined to learning the prevailing values, norms, and behavioural patterns. Socialization can take place through mentorship programmes, training and orientation, and reward systems. Other steps that an organization can take to socialize employees include: Offering them interesting and challenging work. Providing them proper and relevant training. Providing them with objective and timely feedback. Appointing an experienced supervisor to conduct the socialization process .Designing an informal orientation programme. Assigning new employees to work groups that are highly satisfied and have high morale. Socialization is also necessary when organizational member move up the hierarchy. Socialization can also take place through job rotation.
- **Holland's Typology of Matching Personalities with Jobs:** John Holland established a relationship between personality characteristics, the requirements of a job, and job performance. Developed a vocational preference inventory and plotted the results in the form of a hexagonal diagram, with each corner representing an occupational personality type. The closer two personality types are in the hexagon, the more compatible they are. Those that are adjacent to each other are similar,

whereas those that are diagonally opposite are dissimilar in nature. When personality type and occupation match each other, job satisfaction is high and turnover is low.

Type Personality	Personality Characteristics	Congruent Occupations
Realistic: Prefer physical activities that require skill, strength, and coordination	Shy, genuine, persistent, stable, conforming, practical	Mechanic, Drill press operator, assembly line worker, farmer
Investigative: Prefers activities that involve thinking, organizing and understanding	Analytical, original, curious, independent	Biologists, economists mathematician, news reporter
Social: Prefers activities, that involve helping and developing others	Sociable, friendly cooperative, understanding	Social worker psychologist, teacher counselor, clinical
Conventional: Prefers rule regulated, orderly, and unambiguous activities	Conforming, efficient, practical, unimaginative, inflexible	Accountant, Corporate manager, bank teller, file clerk
Conventional: Prefers · ruleregulated, orderly, and unambiguous activities influence others and attain power	Self-confident, ambitious, energetic, domineering	Lawyer, real estate agent, PR specialist, small business manager
Artistic: Prefers ambiguous Unsystematic. Activities that allow creative expression	Imaginative, disorderly, idealistic, emotional, impractical	Painter, musician, writer, interior decorator

CHAPTER 2
BODY LANGUAGE

The body sends a continuous flow of hints and cues. Even, if a person stops talking, 'he cannot stop communicating through body idiom; he must say either the right thing or the wrong thing. He cannot say nothing.' Presidents are not the only people who can communicate a lot while saying little. We all use 'body English.' We lean; we point; we look away. Our hands move to punctuate our sentences. We frown, smile, fold our arms, move forward, then move away. We express our liking of another person, for example, by looking at them attentively, asking questions and leaning forward, while with someone we dislike we remain silent, look away and lean back. We tell somebody about our social or organisational status by walking tall or hunching down, relaxing or being tense, gazing steadily or blinking frequently.

COMMUNICATIVE FUNCTIONS OF THE BODY

- **Body Shape and Appearance:** The body conveys meaning even when it is not in motion. For example, although no proven connection exists between physique and behaviour, people tend to expect such connections and base decisions on them, whether they realize it or not. Tall people are likely to be considered more credible than shorter people; accordingly, one study found that taller employees tend to receive higher salaries; even when their qualifications are equivalent to those of shorter employees. Skin colour and gender also affect communication. For example, members of upper class, predominantly white communities, perceive whites as more credible than African Americans. Men also tend to be seen as more credible than women, even by women. Although such findings vary according to cultural norms, they show that physical attributes can have a significant impact on people's attempts at communication. Dress is also a form of non-verbal communication. It would be impossible to list all the things that we can communicate through our clothing; the list would vary greatly with time, place and situation. We can only say that skillful communicators attend to their apparel. When a speaker presents himself, we see him before we start hearing him. Immediately, we begin developing impressions of his abilities and attitudes based on the non-verbal signals he sends. This is why body language is so critical in oral communication. Body language includes every aspect of your appearance, from what you wear, how you stand, look and move, to your facial expressions and physical habits such as nodding your head, jingling change in your pocket or pulling your necktie. Your use of space and gestures are other key concerns.

- **Personal Appearance:** Personal appearance plays a significant role; people see you before they hear you. As you adapt your language to an audience, you should also dress accordingly. Appearance includes clothes, hair, jewelry, cosmetics and so on. In today's society, the purpose of clothing has changed from fulfilling a basic need to making a personalized statement. Clothes accent the body's movements. You should expect appearance to be a part of the message you communicate. Plan your appearance in such a way that it communicates effectively to others. Even before the speaker utters his first syllable, you begin to form an opinion about him and visualize the way he is going to talk. One's appearance may put the audience into a resistant or hostile attitude or induce in them a receptive mood. You should be clean and well groomed, conforming to the need of the occasion. Appearances communicate how we feel about ourselves and how we want to be viewed.
- **Eye Contact**
  a. It is rightly said that one who has the ability to read the eyes can read the thoughts.
  b. Eye contact is one of the most important and powerful parts of facial expression.
  c. Direct eye contact builds trust and confidence between speaker and receiver.

d. It builds emotional relationship.

e. Prolonged eye contact signals admiration, belief and can build anxiety.

f. Averted eyes show anger, hurt feelings etc.

g. Eye gaze shows power and dominance.

h. Eyes are considered the window to the soul.

i. You look at the eyes of a speaker to help find out the truthfulness of his speech, his intelligence, attitudes and feelings.

j. Eye contact is a direct and powerful form of non-verbal communication.

k. You use your eyes to collect information.

l. Looking directly at listeners builds rapport.

m. It tells the audience that you are sincere in what you say and you want them to pay attention.

n. Prolonging the eye contact for three to five seconds, however, gives the impression of staring.

o. Eye contact is especially important when you start the presentation and at the beginning of each new section.

- **Gesture:** Gestures play an important role in non-verbal communication. The language of gestures is associated with hand and arms movement with entire body capable of showing gestures.
- **Symbols:** With relatively fixed verbal translation, a large number of body movements act as symbols.

  E.g.

  a. First finger on the lips indicates, 'To keep quiet',

  b. Yawning indicates 'I am bored' or 'I am angry' or 'I have no interest',

  c. Nodding the head up and down indicates 'Yes' or 'agree',

  d. Showing first and second finger in V shape indicates victory,

  e. Tapping finger against skull indicates 'I am thinking'.

a. **Illustrators:** The movements of hands and arms go simultaneously with the words of speaker. e.g. when we say the national anthem, we end it with "Jai Hind" and at the same moment, we raise our right hand in air.

b. **Regulators:** Regulators control oral communication by alerting the sender to the need to hurry up, go slow, turn left or right etc. Regulators include smiles, head movement, shift in eye contact, shoulder slaps etc.

c. **Displays:** It indicates emotional states such as anger or embarrassment usually occurring in facial expression.

d. **Adaptors:** These body gestures vary from person to person. These body gestures are automatic and are not planned. Gesture is the movement made by hands, arms, shoulders, head and torso. Skillful gestures can add to the impact of a speech. A well-timed gesture not only drives a point home but also enhances the impact and adds greater value to what is being said. Similarly, an ungainly gesture (like playing with your key chain or button) can mar the effectiveness of the message.

e. **Haptic:** It is the study of touch. It ranges from touch to cuddling, hugging, hand shaking, and even kicking.

f. **Proxemics:** It is the study of space distancing. It refers to the distance generally we maintain in different situations. E.g. Intimate: zero, personal: 1.5 ft-4 feet, social: 4.

- **Posture:** A person's general posture indicates some meaning without any specific gestures. Body posture conveys variety of messages. Postures means the position of body or body parts i.e. the way you sit, the way you walk the way in which you stretch your legs. In organisational life postures and gestures play an important role. In corporate sectors persons postures and gestures can tell much about how effectively the person will perform in an organisation. Postures include side-to-side movement, forward and backward movements and vertical movements of arms, hands and body. Posture generally refers to the way we hold ourselves when we stand, sit or walk. Novice speakers are unsure of what to do with their body. Certain mannerisms creep in, usually out of nervousness pacing constantly, bobbing the shoulders, fidgeting with notes, jingling coins, either constantly moving or remaining static. These mannerisms show his nervousness to the listeners and reduce the final effect of his speech or presentation on them. Standing, sitting or walking in a relaxed way is a positive posture, which will encourage questions and discussion. Squarely facing an audience, being comfortably upright and evenly distributing your weight are all aspects of posture that communicate professionalism, confidence, attention to detail and organisation. Be and appear at ease; neither remains static nor fidget. Move occasionally to hold attention, suggest transitions and increase emphasis.

a. Slumped posture- indicates low spirits.
b. Erect posture- indicates high spirits, energy and confidence.
c. Lean forward- indicates open, honest and interested.
d. Lean backward- indicates defensive or disinterested.
e. Crossed arms- indicates defensive and not ready to listen.
f. Uncrossed arms- indicates willingness to listen.

- **Body Movements:** On the regular basis, people should know what other people need to understand and what other people are communicating with their posture and gestures.

  (a) Attitude (b) Status (c) States and Moods

  a. The movement of legs and shoulders conveys the attitude of communicators.
  b. Status influences body movement.
  c. One can make out easily the status of manager and worker by observing their body language.
  d. States of moods depends upon the emotional level of an individual; anger, happiness, sorrow etc. can easily be made out from an individual's body language.

- **Facial Expression:** Along with posture and gesture, facial expressions also play an important poet. The face is the most expressive part of your body. The face is probably the single most important source of non-verbal communication. The muscles within the face allow it to convey several emotions simultaneously. Faces are sometimes hard to read. A smile stands for friendliness, a frown for discontent, raised eyebrows for disbelief, tightened jaw muscles for antagonism, etc. Facial expressions are subtle. You can use them in a variety of ways: to aid, inhibit, or complement your communication. The face rarely sends a single message at a time. Instead, it sends a series of messages - your facial expressions may show anxiety, recognition, hesitation and pleasure in quick succession, racial expression is difficult to interpret. People tend to hide their true feelings and reveal only those expressions required at a particular movement at times.

The five basic expressions are:

a. Inhibited-restricted and solid

b. Uninhibited-spontaneous and impetuous

c. Substitute-happy with a long face

d. Frozen-no change in expression

e. Blank-no expression at all

CHAPTER 3

PREPARATION OF SELF INTRODUCTION

A self introduction speech is often called for at a first meeting of a group. It could be a work based seminar, a hobby group, your new class at the start of the term. The one thing these settings generally share is the need to make a bunch of strangers feel at ease - to quickly establish a sense of unity or belonging. One of commonest ways of breaking the ice is to have each member of the group give a brief self introduction speech.

EIGHT STEP TO SELF INTRODUCTION

- **Stating your name clearly**
- **Placing yourself:** Where you are from, the organization you belong to, the position you currently hold.
- **Background:** What can you share that is related to the group's core purpose for meeting? Is it an event, experience, a particular skill or educational qualification?
- **Interest, passion or goal:** What particularly interests you? What drives you? What is the personal goal you want to achieve within this group?
- **Sharing personal details:** Hobbies or pet peeves
- **Offer an example of one of your accomplishments:** Don't just tell someone you have a skill or expertise, prove it to them.
- **What are you looking for:** This is the most important part of networking, and often the hardest part for most people. You have to ask for assistance. Most people really want to help you, but they can not help you if they have no idea of what you need. Whether you are looking for a new employer,

networking contacts to expand your client base, or just to make connections with industry professionals and expand your knowledge, be sure to tell your listener what you need.

- **Unity:** What do you share in common with someone else in the group?

CHAPTER 4
COMMUNICATION SKILLS

Communication is a word derived from the Latin word communis, which means 'to make common' or 'to share'. Communication is the act of conveying intended meaning to another person through the use of mutually understood signs and language. Communication is the art of transmitting information, ideas and attitudes from one person to another.

- **Theodorson (1969):** Communication as the transmission of information , ideas, attitudes, or emotions from one person or group to another or others primarily through symbols.
- **Osgood *et al.* (1957):** There is communication wherever one system or source influences another, by manipulation of alternative symbols transmitted over the channel.
- **Gerber (1957):** Communication as social interaction through messages.

CHARACTERISTICS OF COMMUNICATION

- Communication is a process: Communication is a two way process which involves; listening to others (Receiving) message Asserting/Expressing (Sending).
- Communication is a dynamic: it is ever changing depending on the variables at play.
- Communication is a complex process.
- Communication is a two-way process of reaching mutual understanding, in which participants not only exchange (encode-decode) information but also create and share meaning.
- Communication involves the sharing of information using a code.

- Communication occurs between people and sometimes animals.
- Communication is irreversible: once one has communicated something it cannot be recalled back.
- Communication is a system.
- Communication must have the elements of communication: Source, receiver, channel, message, noise, feedback.
- Communication can be verbal/ nonverbal or visual.
- Communication can be accidental especially in non-verbal.

FUNCTIONS OF COMMUNICATION

Human beings communicate for various reasons. Here are some of the reasons why we must communicate:

- To change in behavior
- To influence others
- To express our thoughts and emotions through words & actions
- It is a tool for controlling and motivating people
- It is a social and emotional process
- Communication for improving self-confidence
- Entertain
- Educate
- Establish relationships
- Inform
- Solve problems
- Make orders
- Give directions

LEVELS OF COMMUNICATION

- Intrapersonal (Within a person)
- Interpersonal (Face to face)
- Group communication (Between or within group)
- Mass communication
- Inter country/ Development

THE COMMUNICATION EQUATION

- **What you hear:**
  a. Tone of voice
  b. Vocal clarity
  c. Verbal expressiveness

  } 40 % of the message
- **What you see or feel**
  a. Facial expression
  b. Dress and grooming
  c. Posture
  d. Eye contact
  e. Touch
  f. Gesture

  } 50 % of the message
- **Words:** 10 % of the message

THE 7 Cs OF COMMUNICATION

- **Clear:** When writing or speaking to someone, be clear about your goal or message. What is your purpose in communicating with this person? If you're not sure, then your audience won't be sure either. To be clear, try to minimize the number of ideas in each sentence. Make sure that it's easy for your reader to understand your meaning. People shouldn't have to "read between the lines" and make assumptions on their own to understand what you're trying to say.
- **Concise:** When you are concise in your communication, you stick to the point and keep it brief. Your audience doesn't want to read six sentences when you could communicate your message in three.
- **Concrete:** When your message is concrete, then your audience has a clear picture of what you're telling them. There are details (but not too many!) and vivid facts. Your message is solid.
- **Correct:** When your communication is correct, it fits your audience. And correct communication is also error free communication.

- **Coherent:** When your communication is coherent, it's logical. All points are connected and relevant to the main topic, and the tone and flow of the text is consistent.
- **Complete:** In a complete message, the audience has everything they need to be informed and, if applicable, take action. Does your message include a "call to action", so that your audience clearly knows what you want them to do? Have you included all relevant information – contact names, dates, times, locations, and so on?
- **Courteous/consideration:** Courteous communication is friendly, open, and honest and does not illicit emotions. There are no hidden insults or passive aggressive tones. You keep your reader's viewpoint in mind, and you're empathetic to their needs. You must always put yourself in the shoes of the person you are talking to and ask yourself how you would feel if you were to be addressed the way you are addressing your receiver. Consideration in communication creates a healthy work environment.

COMMUNICATION SKILLS

Importance of Communication Skills

- Communication skills not just handy; they are critical tools for success
- Important for survival in real world environment
- Communication skills are the building blocks of academic life and career,
- Communication skills occupies central positioning a students' life and that of practicing professional;
- Basic skills for survival at university, higher learning.
- Useful in training of professionals and acquisition of:
- Cognitive skills
- Psychomotor –(Technical) skills
- Social –(Comm. skills)

Listening skills

"We were given two ears but only one mouth, because listening is twice as hard as talking."

"If we were supposed to talk more than listen, we would have been given two mouths and one ear."

—Mark Twain

Listeners must first hear what is said. Listening skills involve identifying and selecting relevant points recognized as having meaning; that are understood and held in short-term memory. These can be related to what has gone before and to what comes after. Any information considered important is selected and stored for future reference in the long term memory. Decoding (understanding) a message is generally easier for the listener if a person is speaking rather than reading something out loud. In addition the speaker's facial expressions, and the stress placed on words help the listener to understand the message.

Listening skills

- Active Listening
- Responding
- Paraphrasing
- Asking questions for clarification
- Mirroring the other person's language
- Attentive Body Language
- Following
- Reflecting
- Empathy Blockers

Developing effective listening skills involves two specific steps (Hartley & Bruckman, 2002).

a. To develop the ability to recognize and deal with barriers that prevents you listening with full attention.

b. To develop and use behaviors which help you to listen. Such behaviors can also serve to let the other person know that you are giving them your full attention.

Active listening is composed of six distinct components

a. **Hearing:** The physiological process of receiving sound and/or other stimuli.

b. **Attending:** The conscious and unconscious process of focusing attention on external stimuli.

c. **Interpreting:** The process of decoding the symbols or behavior attended to.

d. **Evaluating:** The process of deciding the value of the information to the receiver.

e. **Remembering:** The process of placing the appropriate information into short-term or long-term storage.

f. **Responding:** The process of giving feedback to the source and/or other receivers.

Improve listening skills

Listening effectively is hearing and understanding what a speaker is saying and how it applies to you, and then remembering it for future use and evaluation. There are ways to improve your listening skills for lectures. The following is a list of some basic techniques:

- Recognize how ideas are organized. Lectures usually begin with some type of introduction, followed by a thesis statement which is supported by additional information. Most professors bring closure to their lecture by summarizing what they have covered. Learn to identify the lecture style that is used by your professor.
- Become involved in what is being said. Be an avid listener. Constantly analyze what is being said.
- Cut through (or screen out) distractions.
  a. background noise
  b. unusual accents, dialects, and language mistakes
  c. speaker disorganization, emotion, or habits
  d. unrelated material
  e. your own inner voice
- Organize statements into main points and supporting reasons. Using an outline form may be helpful.
- Discriminate between relevancies and irrelevancies. Remember that not all information is important.
- Maintain an active body state. Keeping alert and having eye-contact with the speaker will help you listen more effectively.

- Find area of interest - maintain eye contact even if the subject appears boring. There will be some information that will be useful.
- Judge the content, not delivery. Find out what the speaker knows, not how he presents it.
- Withhold evaluation until comprehension is complete – don't become preoccupied with rebuttal before idea is completely presented. Don't listen defensively.
- Listen for ideas - main ideas, central facts, organizational patterns.
- Work at listening - spend energy to give conscious attention.
- Resist distractions - a matter of concentration
- Exercise your mind - develop an appetite for hearing a variety of presentations difficult enough to challenge your mental capacities.
- Keep your mind open - be careful of emotional impact of certain words - don't listen defensively composing a rebuttal.
- Capitalize on thought speed.
    a. Speech speed - 100-200 words per minute
    b. Thought speed - 400-500 or more words per minute.

Writing Skills

We learn to read by reading, and
We learn to write by writing.
But
We also learn to read by writing, and
We learn to write by reading.

Writing is:

- A process of putting thought ,ideas and concepts into words using continuous prose;
- A thinking process;
- A means of building fluency;
- A way of developing accuracy (in grammar, vocabulary, etc.);
- A critical skill for academic or professional success
- Is a way to demonstrate proficiency

- Helps us discover what we do or do not know
- Is more than a paragraph or essay

Writing Skills

- Ability to put sounds down on graphic form according to the conventional sound-spelling rules.
- Ability to spell English words correctly, including using correct punctuation and capitalization.
- Ability to do writing practice, such as dictation, grammar exercises, constructing dialogues according to the model, simple translation exercises.
- Ability to write short compositions, including functional writing skills, such as writing simple letters, taking notes, writing outlines and summaries.

Writing needs to have:

- Form (letter, email, summary, report, etc.)
- Audience
- Topic
- Purpose (describe, explain, persuade, etc.)
- Clear
- Grammatically correct
- Have variety
- Precise
- Interesting pace
- Not be emotional
- Appropriate: language and audience
- Supported by evidence

The Process of Writing

Planning

Involves gathering ideas and focusing on the topic: Topic selection, gathering ideas, extensive reading and interpretation. When planning one must ask himself/hers

- What is the purpose of my writing?
- Who is my audience?
- What are the special requirements of my writing task?

Shaping

Considering how best to organize your ideas on pen and paper. Shaping involves:

- Determining the tone
- Drafting a thesis statement
- Outlining ideas
    - Drafting
    - Revising
    - Editing
    - Proof reading

Speaking skills

Speaking:

Speaking is an act of making vocal sounds. We can say that speaking means to converse, or expressing one's thoughts and feelings in spoken language. To speak often implies conveying information. It may be from an informal remark to a scholarly presentation to a formal address.

Speaking skills:

- Speaking skills are the skills that give us the ability to communicate effectively.
- These skills allow the speaker, to convey his message in a passionate, thoughtful, and convincing manner.
- Speaking skills also help to assure that one won't be misunderstood by those who are listening.

What to do Before Speaking?

The first step to speaking is to think and the first step to perfect speaking is to think positive. Good speakers have the ability to quickly analyze and absorb the information given to them, assess it fast and to make a decision and communicate that decision to other. Keep in mind to:

- **Be an active listener:** The key ingredient towards making you a person who can think on the spot and respond intelligently is to be an active listener. This means listening carefully and giving your full attention to the words, tone, emotion and logic behind what the other one is saying.

- **Be a quick organiser of thoughts:** All of us have the ability to think fast. The trick is to adopt some frameworks or models to structure new information into something coherent that we can respond with. One can break down issues or ideas into past, present and future.
- **Structure your speech in your mind:** These are useful frameworks to organised ideas and thoughts quickly.

Speaking Skills:

- **Be prepared and practice:** The more you know what you want to say, the better you'll get at it. First, brainstorm the topic of your speech and research it, if you need to. Write down all the points you want to make and then organize them into an outline. Then, practice your speech out loud at least 3 to 5 times.
- **Know your audience:** The better you know your audience or listeners, the easier it will be to connect to them as you speak. When you are able to make that connection, you'll hold their attention.
- **Pay attention to the old KISS principle, that is, Keep It Short and Simple:** When you are speaking don't beat around the bush or try to impress with complex metaphors. Stories, however, can be a powerful public speaking tool, especially when they contain at least a hint of humor. But again, keep them short and on point. Shorter messages leave more impression. They're also easier for your audience to remember.
- **Interact with your audience:** Lectures will rarely have the same impact on an audience that an open discussion will. Look for opportunities to involve your audience in what you are speaking about. Ask for validation of points you are making ("Am I right?" "Has that ever happened to you?") Or allow time for questions. Also, make sure to establish eye contact with your audience, and keep it throughout your speech.
- **Speak with sincerity and passion:** When a person wants to leave a lasting impression with the audience about one's topic then be sure that you are true to yourself and your topic as you

speak. Don't be afraid to inject enthusiasm and passion into the speech as well.

- **Close your speech in a memorable way:** Give your audience something to think about as you finish up your speech. Certainly, it's a good idea to summarize your main points one more time, but then finish up with an inspiring story or quote, or leave them with a thought provoking question.
- **Fluency:** The main goal is fluency. Remember that one don't have to know many complex grammatical structures to achieve that goal! First of all try to speak as fluent as possible (even making some grammar mistakes). Then, after making one's speaking fluent, one can focus on grammar aspects.

Improve Speaking Skills

- Practice where you can, when you can: Any practice is good whether you speak to someone who is a native English speaker or not.
- It's important to build your confidence. If possible, use simple English sentence structure that you know is correct, so that you can concentrate on getting your message across.
- Try to experiment with the English you know. Use words and phrases you know in new situations. Native English speakers are more likely to correct you if you use the wrong word than if you use the wrong grammar. Experimenting with vocabulary is a really good way of getting feedback.
- Try to respond to what people say to you. You can often get clues to what people think by looking at their body language. Respond to them in a natural way.
- Try NOT to translate into and from your own language. This takes too much time and will make you more hesitant.
- If you forget a word do what native English speakers do all the time, and say things that 'fill' the conversation. This is better than keeping completely silent. Try using um, or err, if you forget the word.
- Don't speak too fast! It's important to use a natural rhythm when speaking English, but if you speak too fast it will be difficult for people to understand you.

- Try to relax when you speak you'll find your mouth does most of the pronunciation work for you. When you speak English at normal speed, you'll discover that many of the pronunciation skills, such as linking between words, will happen automatically.
- Remember, when speaking English try to become less hesitant and more confident. Don't be shy to speak the more you do it, the more confident you'll become. Remember to be polite use "please" and "thank you" if you ask someone to do something for you.

Overcoming Barriers for Effective Communication

Physical barriers

- **Competing stimulus:** Another conversation (within hearing distance), loud music, traffic noise (in the background), crows cawing, plane overhead - can drown message.
- **Environmental stress:** High temperature plus humidity, poor ventilation, vibrations felt, strong glare can contribute to distractions (in sending and receiving messages).
- **Subjective stress:** Sleeplessness, ill health, effects of drugs, mood variations give rise to stress, leading to difficulty in listening and interpretation.
- **Ignorance of medium:** The various media for communication are oral, written, audio, visual, audio-visual. Use of a medium, which the communicator is not familiar with could turn the medium itself into barrier (e.g. maps, charts used to instruct workers who have not been taught to read maps will alienate workers immediately).

Psychological barriers

- **Self image:** Enmeshed in the 'frame of reference' is the self-image or self-concept of a person. Owing to this self-image people establish their point of view, and interpret messages in accordance with their self concepts. They interpret 'reality' in the light of these pre-conceived notions. As a part of this system, people tend to listen and interpret favorably towards those messages, which give a further their self-image and reject those messages, which threaten that image.

- **Resistance to change:** "The risk of being changed is one of the most frightening prospects many of us can face"(On Becoming a Person by Carl Rogers, www.listeningway.com). It is a human tendency to resist change and new ideas. However, the effective communicator does not wait for resistance to accumulate but takes people into confidence, even at planning stage, lends a ear to their point of view, involves them in change, converses about benefits, assures them security will not be affected and explains why change is necessary.
- **Defensiveness and fear:** This is strongly allied to the barrier raised by a 'resistance to change'. One of man's most compelling needs is to justify himself. "Fear is an effect of great potency in determining what the individual will perceive, think and do", said Izard and Tomkins. Together with connected emotions of nervousness, anxiety and tension, fear is the most restricting of all effects, resulting often in 'tunnel vision'. It also gives slow, narrow thinking, which selects and distorts communication. Appropriate examples of this are interviews and exams. Some psychologists propose that a little anxiety is good; it heightens attention, improves performance, releases certain hormones and facilitates learning by a greater nerve messages to brain. In other words, fear and anxiety can be turned into a source of energy and confidence

Interpersonal barriers

Withdrawal is an absence of interpersonal contact. It is both refusals to be in touch with others.

Linguistic and cultural barriers

Language is an expression of thoughts/expression of people in terms of cultural environment. When same language is used in different cultures, it takes another colour. Language can be appropriated to suit varied situations e.g. financial terms, medical terms, psychological terms etc. Language facilitates understanding, but it can also prove to be a barrier to communication. A language, whether verbal or nonverbal, is ambiguous in nature. Words are mere symbols. Symbols are comprehended differently by participants in communication. Cultural differences have an impact on language and meanings. Cultures provide people with ways of thinking, ways of seeing, hearing, and interpreting the world. Thus, the same words can possess myriad meanings for who speak the same language, but are native to different cultures. When languages are distinct, communication is carried out through translation, which increases the probability of misunderstandings.

Mechanical barriers

Mechanical barriers are those raised by channels employed for interpersonal or mass communication. Channels become barriers when the message is impeded by disturbances. In such a case, it a) increases the difficulty in reception or b) prevents some elements of the message reaching its destination or both. Absence of communication facilities can also be mechanical barrier. Technically such barriers are clubbed as 'channel noise'. These include any disturbance, which interferes with the fidelity of physical transmission of a message. Some examples of mechanical barriers are static on radio, smeared ink on newspaper, a rolling screen on TV, a barely readable point size or a film projector, video/DVD that does not function properly. Barriers or channel noise can be associated with cultural or social issues, language, customs, beliefs, motives, aspirations, assumptions across geographic lines or simply illiteracy. A large number of sources of 'noise' or interference can infiltrate the channel of communication. This may occur when people are not very familiar with each other and they do not understand the sources of error. This kind of interference is frequent in a work setting where the people have not known each other very long. This may lead to conflicting relationship, which can make the process of communication even more complex.

Stereotypes

Stereotypes are widely circulated ideas or assumptions about particular groups. Stereotypes are usually negative attitudes which people use to justify discrimination of conflict against others. Stereotyping acts as a barrier to communication because people make pre-conceived judgment about people which are unfounded if their character does not relate to their appearance. Stereotyping has a halo effect. Halo effect is the use of a single attribute to describe a person or object fully. For example, if a person is friendly we may use this attribute to assume that they are punctual and good at their job. Another important aspect of stereotyping is perception.

Authority

Barriers to communication for one reason or the other often get neglected. Knowing them is synonymous to knowing about one's own barriers sprouting out of one's personality. Barriers to communication can lead to misunderstanding and confusion.

CHAPTER 5

BUILDING SELF-ESTEEM AND SELF-CONFIDENCE

SELF-ESTEEM

- Self-esteem is a realistic respect for or favorable impression of one self; Self-respect. (Dictionary.com)
- Self-esteem refers to the way we see and think about ourselves. It is a state of mind.
- Self-Esteem is a combination of self-image, ideal self, and Pygmalion-self.
- Having high self-esteem means having feelings of confidence, worthiness and positive regard for yourself.
- People with high self-esteem feel good about themselves. They feel a sense of belonging and security. They respect themselves and appreciate others.

WHY DEVELOP SELF-ESTEEM?

The rewards of developing self-esteem include being able to take risks, having positive relationships, not being held back by fears and insecurities, pursuing your dreams and desires, making good choices and reaching your goals.

Signs of Positive Self-esteem

There are some outword signs of positive self-esteem as discussed below:

- Confidence
- Self-direction
- Non-blaming behavior
- An awareness of personal strengths

- An ability to make mistakes and learn from them
- An ability to accept mistakes from others
- Optimism
- An ability to solve problems
- An independent and cooperative attitude
- Feeling comfortable with a wide range of emotions
- An ability to trust others
- A good sense of personal limitations
- Good self-care

WHAT IS LOW SELF-ESTEEM?

Low self-esteem is a debilitating condition that keeps individuals from realizing their full potential. A person with low self-esteem feels unworthy, incapable, and incompetent. In fact, because the person with low self-esteem feels so poorly about him or herself, these feelings may actually cause the person's continued low self-esteem.

Here are Some Signs of Low Self-esteem

- Negative view of life
- Perfectionist attitude
- Mistrusting others – even those who show signs of affection
- Blaming behavior
- Fear of taking risks
- Feelings of being unloved and unlovable
- Dependence – letting others make decisions
- Fear of being ridiculed.

Building Self-esteem

- Learn to be more assertive and not feel guilty about saying no
- Think positively
- Believe on your self
- Make plans for the future
- Set and achieve goals

- Do something challenging each day
- Look your best
- Do something for someone else
- Learn a new skill
- Handle things one at a time
- Use criticism constructively
- Ask for help – take advantage of learning opportunities
- Allow personal growth time each day
- Reward yourself
- List your accomplishments each evening
- Choose friends who value and respect you
- Focus on positive aspects about yourself.
- Replace negative self-talk with supportive self-talk.
- Act in accordance with your values.
- Consider mistakes learning opportunities.
- Try new activities to discover your talents.
- Exercise regularly to feel more energized.
- Focus on the positive aspects about yourself
- Supportive self-talk
- Work toward accomplishments rather than perfection
- Try new activities
- Exercise regularly

Self-Confidence

- Self-confidence is the difference between feeling unstoppable and feeling scared.
- Self-confidence is a term used to describe how secure a person is in his own decisions and actions & how free he is from doubts.
- Doing what you believe to be right, even if others mock or criticize you for it.

Building Self Confidence

- Groom yourself
- Dress nicely
- Photoshop your self-image
- Think positive
- Kill negative thoughts
- Get to know yourself
- Act positive
- Be kind and generous
- Get prepared
- Know your principles and live them
- Speak slowly
- Stand tall
- Increase competence
- Set a small goal and achieve it
- Change a small habit
- Focus on solutions
- Smile
- Volunteer
- Be grateful
- Exercise
- Empower yourself with knowledge
- Do something you've been procrastinating on
- Get active
- Work on small things
- Clear your desk

CHAPTER 6
ATTITUDE

INTRODUCTION

Attitudes involve some knowledge of a situation. However, the essential aspect of the attitude is found in the fact that some characteristic feeling or emotion is experienced and, as we would accordingly expect, some definite tendency to action is associated. Subjectively, then, the important factor is the feeling or emotion. Objectively it is the response, or at least the tendency to respond. Attitudes it is the response, or at least the tendency to respond. Attitudes are important determinants of behaviour. If we are to change them we must change the emotional components. All port has defined attitude as a mental and neutral state of readiness organised through experience, exerting a directive or dynamic influence upon the individual's response to all objects with which it is related. Attitude is a psychological construct, it is a mental and emotional entity that inheres in, or characterizes a person. They are complex and an acquired state through experiences. It is an individual's predisposed state of mind regarding a value and it is precipitated through a responsive expression toward a person, place, thing, or event (the attitude object) which in turn influences the individual's thought and action. It is undeniable that the concept of "attitude" has become something of a factotum for both psychologists and sociologists. But, in spite of all the animadversions of critics, the term is now in nearly universal use and plays a central role in most of the recent systematic, studies in social psychology. It is therefore a concept which students must examine with unusual care.

- Attitude can be formed from a person's past and present.
- An attitude can be as a positive or negative evaluation of people, objects, events, activities, and ideas.
- An attitude is an evaluation of an attitude object, ranging from extremely negative to extremely positive.

- Attitude is a favorable or unfavorable evaluative reaction toward something or someone, exhibited in ones beliefs, feelings, or intended behavior **(Myers, p. 36).**
- Attitude defined as a mental or neural state of readiness, organized through experience, exerting a directive or dynamic influence on the individual's response to all objects and situations to which it is related. A simpler definition of attitude is a mindset or a tendency to act in a particular way due to both an individual's experience and temperament **(Allport,1935)**.
- An attitude is readiness for attention or action of a definite sort **(Baldwin, 1901-1905)**.
- Attitudes are literally mental postures, guides for conduct to which each new experience is referred before a response is made **(Morgan, 1934)**.
- Attitude are the specific mental disposition toward an incoming (or arising) experience, whereby that experience is modified, or, a condition of readiness for a certain type of activity **(Dictionary of Psychology, Warren, 1934)**.
- An attitude is a complex of feelings, desires, fears, convictions, prejudices or other tendencies that have given a set or readiness to act to a person because of varied experiences **(Chave, 1928)**.
- Attitude a more or less permanently enduring state of readiness of mental organization which predisposes an individual to react in a characteristic way to any object or situation with which it is related **(Cantril, 1934)**.
- An attitude is a tendency to act towards or against something in the environment which becomes thereby a positive or negative value **(Bogardus, 1931)**.
- An attitude is a mental disposition of the human individual to act for or against a definite object **(Droba, 1933)**.
- An attitude denotes the general set of the organism as a whole toward an object or situation which calls for adjustment **(Lundberg, 1929)**.
- Attitudes are modes of emotional regard for objects, and motor "sets" or slight, tentative reactions toward them **(Ewer, 1929)**.

HISTORY OF THE CONCEPT OF ATTITUDE

Like most abstract terms in the English language, attitude has more than one meaning. Derived from the **Latin '*aptus*'** it has on the one hand the significance of "fitness" or "adaptedness," and like its by-form *aptitude* connotes a subjective or mental state of preparation for action. Through its use in the field of art, however, the term came to have a quite independent meaning; it referred to the outward or visible posture (the bodily position) of a figure in statuary or painting. The first meaning is clearly preserved in modern psychology in what are often referred to as "mental attitudes"; and the second meaning in "motor attitudes." Since mentalistic psychology historically precedes response psychology, it is only natural to find that mental attitudes are given recognition earlier than motor attitudes. One of the earliest psychologists to employ the term was Herbert Spencer.

Following closely in the same vein of thought, Faris (1925) proposed additional refinements. He would distinguish between conscious and unconscious attitudes, between mental and motor attitudes, between individual and group attitudes, and between latent and kinetic attitudes. Park (see Young, 1931), who is likewise in essential agreement with this school of thought, suggests four criteria for an attitude:

- It must have definite orientation in the world of objects (or values), and in this respect differ from simple and conditioned reflexes.
- It must not be an altogether automatic and routine type of conduct, but must display some tension even when latent.
- It varies in intensity, sometimes being predominant, sometimes relatively ineffective.
- It is rooted in experience, and therefore is not simply a social instinct.

COMPONENTS OF ATTITUDES

- **Cognitive:** Our thoughts, beliefs, and ideas about something. When a human being is the object of an attitude, the cognitive component is frequently a stereotype, e.g. "welfare recipients are lazy"

- **Affective:** Feelings or emotions that something evokes. e.g. fear, sympathy, hate. May dislike welfare recipients.
- **Conative, or behavioral:** Tendency or disposition to act in certain ways toward something. Might want to keep welfare recipients out of our neighborhood. Emphasis is on the tendency to act, not the actual acting; what we intend and what we do may be quite different.

TYPES OF ATTITUDE

- **Assertive:** This is logical, thinking behaviour. It is not driven by your emotions and although it may be natural for a few people, it tends to be learned behaviour. It is about:
  a. Being clear and direct in what say
  b. Stating needs and feelings in a straightforward way
  c. Standing up for rights without violating the rights of others.
- **Aggressive:** Aggressive behaviour is the fight response. Again, this is an in-built programme that can be developed throughout life. People who are aggressive tend to:
  a. Encourage others to do things by flattery or manipulation
  b. Ignore the needs and feelings of others, either intentionally or by default
  c. Take rights for themselves that they don't give to others.
- **Submissive:** Submissive behaviour is the flight response. It is natural behaviour, and depending on upbringing. People who are submissive tend to:
  a. Avoid stating their needs and feelings
  b. Communicate their needs and feelings in an apologetic way
  c. Give others rights that they don't claim for themselves
- **Positive:** A favourable attitude- liking people, objects, situation, etc.
- **Negative:** An unfavorable attitude- does not like people/ objects, etc.
- **Neutral:** Neither favourable nor unfavorable.

THEORIES OF ATTITUDE FORMATION AND CHANGE

- **Functionalist Theory:** Daniel Katz proposed a functionalist theory of attitudes. He takes the view that attitudes are determined by the functions they serve for us. People hold given attitudes because these attitudes help them achieve their basic goals. Katz distinguishes four types of psychological functions that attitudes meet.

  a. **Instrumental:** We develop favorable attitudes towards things that aid or reward us. We want to maximize rewards and minimize penalties. Katz says we develop attitudes that help us meet this goal. We favor political parties that will advance our economic lot - if we are in business, we favor the party that will keep our taxes low, if unemployed we favor one that will increase social welfare benefits. We are more likely to change our attitudes if doing so allows us to fulfill our goals or avoid undesirable consequences.

  b. **Knowledge:** Attitudes provide meaningful, structured environment. In life we seek some degree of order, clarity, and stability in our personal frame of reference. Attitudes help supply us with standards of evaluation. Via such attitudes as stereotypes, we can bring order and clarity to the complexities of human life.

  c. **Value-expressive:** Express basic values, reinforce self-image. Ex: if you view yourself as a Catholic, you can reinforce that image by adopting Catholic beliefs and values. Ex: We may have a self-image of ourselves as an enlightened conservative or a militant radical, and we therefore cultivate attitudes that we believe indicate such a core value.

  d. **Ego-defensive:** Some attitudes serve to protect us from acknowledging basic truths about ourselves or the harsh realities of life. They serve as defense mechanisms. Ex: Those with feelings of inferiority may develop attitude of superiority. Katz's functionalist theory also offers an explanation as to why attitudes change. According to Katz, an attitude changes when it no longer serves its function and the individual feels blocked or frustrated. That is, according to Katz, attitude change is achieved not so much by changing a person's information or perception about an object, but rather by changing the person's underlying motivational and personality needs. Ex: As your social status increases, your attitudes toward your old car may change - you need something that better reflects your new status. (For that matter, your attitudes toward your old friends may change as well).

- **Learning Theory:** (which stresses attitude formation). There are several means by which we learn attitudes.

  a. Classical conditioning: Ex: A father angrily denounces the latest increase in income taxes. A mother happily announces the election of a candidate she worked for. These parents are expressing opinions, but they are also displaying nonverbal behavior that expresses their emotions. For a child watching the parents, the association between the topic and the nonverbal behavior will become obvious if repeated often enough. And the nonverbal behavior will trigger emotional responses in the child: the child feels upset and disturbed when listening to the father and happy when listening to the mother. This is an example of classical conditioning: when two stimuli are repeatedly associated, the child learns to respond to them with a similar emotional reaction. In this case, the stimuli are the attitude topic and the parental emotion. Through repeated association, a formerly neutral stimulus (the attitude topic - taxes or politicians) begins to elicit an emotional reaction (the response) that was previously solicited only by another stimulus (the parental emotion). Whenever tax increases are mentioned, the child feels an unpleasant emotion; when the elected official is mentioned, the child feels a pleasant emotion. Ex: Pavlov's dogs. Bell was rung when dogs received food. Food made dogs salivate. Then whenever a bell was rung, dogs salivated even when food was not present. Ex: When you were a child, parents may have cheered for N.D. football. You may not have even known what N.D. football was, but you liked your parents happy attitude. Now N.D. football evokes that same response in you. Ex: Men with bow ties. Meet a bad man who wears bow ties, and you may come to hate all bow ties.

b. **Instrumental, or operant, conditioning:** Behaviors or attitudes that are followed by positive consequences are reinforced and are more likely to be repeated than are behaviors and attitudes that are followed by negative consequences. Ex: People agree with your opinion.

c. **Observational learning:** Children watch the behavior of people around them and imitate what they see. Ex: If a young girl hears her mother denounce all elected officials as crooks, she may repeat that opinion in class the next day. Whether she continues to repeat that opinion depends on the responses of her classmates, teacher, and parents. That is, observations determine the responses we learn, but reinforcement determines the responses we express.

- **Cognitive Dissonance Theory:** Stresses attitude change - and that behaviors can determine attitudes.

  a. **Cognition** = individuals perception of own attitudes, beliefs, behaviors. *Cognitive dissonance* = feelings of tension that arise when one is simultaneously aware of two inconsistent cognitions. For example, when we act contrary to our attitudes; or, when we make a decision favoring one alternative despite reasons favoring another.

  b. **Consistency** theories hypothesize that, should inconsistencies develop among cognitions, people are motivated to restore harmony.

- **Key Propositions of Dissonance Theory**

  a. Dissonance theory says relationships among two cognitions can be either consonant, dissonant, irrelevant.

  b. Cognitive dissonance is a noxious state. It produces unpleasant physical arousal.

  c. Individual will attempt to reduce or eliminate dissonance- and will try to avoid things that increase dissonance. Ex: Selective observation.

d. Cognitive dissonance can be reduced or eliminated only by (a) adding new cognitions, or (b) changing existing ones. Ex: Can change our minds. Decide we were wrong.

e. **Sources of dissonance**

i. ***Informational inconsistency:*** Receive information that contradicts what they already know or believe. EX: Suppose you believe George Bush did not know about Iran-Contra - and then suppose Oliver North testified that he was the mastermind behind it. (Real life example: some Iranians are said to believe George Bush did head up Iran-Contra, since he used to be head of the CIA and they think the CIA runs the country.)

ii. ***Disconfirmed expectations:*** People prepare themselves for an event that never occurs - or even worse, an event whose opposite occurs. EX: You expect to do well on an exam, and you don't. EX: When prophesy fails. In 1955, Marian Keech predicted that a great flood was going to destroy the Western Hemisphere on Dec. 21. She said she got her information from the planet Clarion. She attracted a band of followers, and received further messages about how the faithful could save themselves. Midnight of the big day came and passed, and nothing happened. At 4:45 a.m., they received a Christmas message informing them that because of their commitment and faithfulness, the earth had been spared.

FUNCTIONS OF ATTITUDE

- Attitudes allow us to **understand the world** around us. Positive attitudes help to come closer to some people and make you agree to their requests. You may identify with role models (e.g. Sachin Tendulkar, Mother Teresa) and try to develop their way of thinking and behaving. The negative attitudes make us stay away from such people or situations.
- Attitudes help **describe the social groups** that each one of us belongs to. As members of a family, group of friends, or, religious, political group we share similar attitudes and this

helps to bind us together. Thus we may favour the religious and political groups we belong to and show unfavorable attitude towards the other groups.

- Attitudes also help us to understand **who we are** or our identity. Attitudes express an individual's values and self-concept. For example, some people value equality while others may value freedom or compassion. Mahatama Gandhi's attitude towards honesty and Mother Teresa's concern for the poor and needy is well known.
- Attitudes help us to **get support, praise and acceptance** from others. People who hold similar attitudes are attracted toward each other. For example, your friend and you share common interests and attitudes, both of you like each other and hence it makes easier to carry on the friendship. Thus attitudes help us to adjust in our social interaction.

RELATIONSHIP BETWEEN ATTITUDE AND BEHAVIOUR

The relationship between attitude and behaviour is complex and it is often difficult to predict one's attitude from his or her behaviour. How many times have you heard people state that they want to be fit but they do not exercise, or that they are worried about global warming but drive a big car which consumes petrol by the gallons! People say one thing but do not necessarily follow it up with appropriate behaviours. Studies have shown that there are certain special conditions under which attitudes can predict behaviour:

- When attitudes are **strong and consistent**, that is when the three components of attitude are clear and stable they better predict behaviour. Weak, unimportant and ambiguous attitudes are less likely to predict behaviour.
- When one holds a strong attitude or when one is **conscious of one's attitude**, thereby it is also **easily recalled** and possible to predict behaviour.
- When attitudes have been formed through **direct experience** prediction of behaviour is more accurate.

- When one acts under social pressure attitude may be expressed in diverse ways. An adolescent may not want to smoke or drink because it affects health. But peer group pressure may force him to drink. Thus his way of thinking is different from his behaviour. When **external influences are minimal** then attitude behaviour relationship is becomes strong.

It is important to note whether behaviour is spontaneous or well planned. When behaviour is deliberated upon and planned, particularly which college to join, or when and whom to marry, these issues become important hence hasty decisions will not be taken. Thus when we have time to think and it is personally relevant then the attitude-behaviour relationship will be strong.

ATTITUDE CHANGE

Well established attitudes tend to be resistant to change, but others may be more amenable to change. Attitudes can be changed by a variety of ways. Some of the ways of attitude change are as follows.

- By obtaining new information from other people and mass media, resulting in changes in cognitive component of a person's attitudes.
- Attitudes may change through direct experience.
- Attitudes may change through legislation.
- Since person's attitudes are anchored in his membership group and reference groups, one way to change the attitude is to modify one or the other.
- Attitude change differs with reference to the situation also.

FACTORS INFLUENCING THE DEVELOPMENT OF ATTITUDES

- **Maturation:** The young child has only a very limited capacity for understanding the world about him and he is consequently incapable of forming attitudes about remote, or complex, or abstract things or problems. At about a mental age of twelve years the child begins to understand abstract terms such as pity and justice, and his capacity for both inductive and deductive reasoning shows a marked and continuous increase during adolescence. As a result of this growth in capacity, he becomes able to understand and react to more abstract and more generalized propositions, ideas and ideals. At the age of four or five years, three characteristics especially deserve mention. These are curiosity, centra-suggestibility, and independence. The child at this age is likely to express his curiosity by asking an endless series of questions. Adolescence is marked especially by the maturation of sex emotions and by the development of altruism and co-cooperativeness. These in large measure furnish the basis for the formation of attitudes that differentiate adults from children. Boys at the age of twelve years may have a distant interest in girls and they may even have crushes on particular girls, but their interest is quite different from what it will be some years later.

- **Physical factors:** Clinical psychologists have generally recognized that physical health and vitality are important factors in determining adjustment, and frequently it has been found that malnutrition or disease or accidents have interfered so seriously with normal development that serious behaviour disturbances have followed.
- **Home influences:** It is generally accepted that attitudes are determined largely by social environment and that home influences are especially important.
- **The social environment:** The home environment is of primary importance in the formation of early attitudes, but friends, associates and the general social environment come to have an increasing influence as the child grows older and has wider social contacts.
- **Government:** The form of the government seems to be an important factor in determining attitudes both towards government itself and towards other things.
- **Movie pictures:** Attendance at movie pictures constitutes another important possible influence in determination of attitudes. Thurston concluded that films definitely change social attitudes, although only about 10 per cent of the attitudes studied seemed to be affected by movie attendance.

- **The teacher:** Brown asked 300 graduate and undergraduate students in educational sociology to evaluate the various factors in their school experience that had been influential in the formation of personality and character traits. According to their judgment, the personalities of their teachers had been the most important single factor, 65.3 per cent thought this influence had been good, but 33.3 per cent thought it had been unfavorable. Only about 10 per cent did not consider, the teacher's influence important.
- **The curriculum:** Thorndike asked 155 teachers to rate eleven subjects and activities on the basis of what they considered the value of these to be or the training of character. Teaching has the highest rank, but athletic sports come next. English literature and history have the best ranks for the regular school subjects; mathematics and foreign languages are ranked much lower. This indicates that, in the opinion of this group of teachers, literature and the social sciences have more influence than other subjects on the determination of attitudes. This seems a reasonable view and it suggests that the units of work and the readings in these areas should be selected with particular reference to their probable influence on the attitudes formed by the students.
- **Teaching methods:** One of the categories in Brown's study was "*manner of presentation*" of subject matter. This was judged to have a favourable effect by 8.0 per cent of the students and an unfavorable effect by 17.7 per cent.

MEASUREMENT OF ATTITUDES

Since the publication of Thurston's procedure for attitude assessment in 1929 (Thurstone & Chave, 1929), employee surveys have been widely used in organizations to obtain information about workers' attitudes toward their environments. This information is helpful for healthcare managers to determine if management is "doing the right things" for retaining and motivating employees. As an example, Lowe, Schellenberg, and Shannon (2003) found that workers who rated their work environments as "healthy" (task content, pay, work hours, career prospects, interpersonal relationships, security) reported higher job satisfaction, morale, and organizational commitment and lower absenteeism and intent to quit. Employee attitude surveys are usually designed using 5-point Likert type ("strongly agree - strongly disagree") or frequency ("never–very often") response formats. Questions typically asked are illustrated in following table. However, as Morrel-Samuel (2002) points out, organizations need to be cautious regarding the design of employee surveys to ensure that problem areas are not overlooked.

Legend (check ✓ the correct number that applies to each question): **1 = Strongly Disagree 2 = Disagree 3 = Neutral** **4 = Agree 5 = Strongly Agree**						
		1	2	3	4	5
1.	Do you feel that your salary reflects your worth to the organization?					
2.	Do you feel appreciated for your work performance?					
3.	Do you feel appreciated for your work performance?					
4.	Are you provided the opportunity for growth/ advancement to higher level tasks?					
5.	Do you feel that you contribute to the overall success of the organization					
6.	Would you recommend the organization to family and friends?					
7.	Are you given the opportunity to learn new skills through formal training?					
8.	Do you feel the organization provides adequate resources to complete your work?					
9.	Do you feel overwhelmed by your workload?					
10.	Do you experience ongoing interests in your job/ tasks?					

CHAPTER 7
LEADERSHIP

LEADER AND LEADERSHIP STYLE

It is difficult for any country to provide enough number of extension workers to reach each and every family for its development programmes. It is rather more difficult for the developing countries where the resources are scarce. This problem can be solved to some extent through the use of local leaders.

A local leader who has adopted improved practices extends the same to others. The common man has much faith in local leaders. A villager would like to hear and imitate his own neighbor as compared to accepting the advises of an outside change agent. Moreover there is a healthy competition among the villagers which promotes action among them.

There is no doubt that extension worker also has to play the role of leader but he may confine his role to the few selected contact farmers. The information from the extension workers will reach these leader farmers which will further trickle down to other people from the leader farmers.

These leaders would like to act as local leaders and feel pleasure in serving to others. They must have certain qualities. He should be a person from the same community and has the same type of resources. It has been often observed that innovators and big farmers do not pass on freely their knowledge to other people in the village due to some gap. It is, therefore, essential for the students of extension education to study the process of leadership. Once the process of dissemination through leaders is known the introduction of new ideas through village leaders can be solved to a great extent.

Leader

- Leader is a person who has been spontaneously considered or chosen as being influential **—O. P. Dahama & O. P. Bhatnagar.**
- Leader is the servant of the group. The position of leader is an essential mechanism of effective group organisation **—Sanderson**.
- Leaders are persons who are selected by the people because of their special interest or fitness to work on some phase of the local programmes. **—J. S. Gang.**
- A leader is one who, in a social situation, can elicit (stimulate) positive reaction from other members of the group. **—B. M. Stogdill.**

Leadership

- According to **Niderfrank (1966)** leadership is essential in simply influencing the attitudes and actions of one or more persons leading towards the achievement of some purpose.
- "*Leadership is an act that influences*" says **Tead (1926).**
- **Cartwright and Zender (1960)** perceived leadership as the performance of acts that help the group to achieve its preferred outcome.
- **Lester (1975)** defines leadership as the resource that an individual or group used to enable the organisation to do what it needs, should or wants to do.
- According to **Davis (1977)** leadership is the ability to persuade others to seek defined objectives enthusiastically.
- **Dahama and Bhatnagar (1985) leadership** is the process who in any social situation with his ideas and actions influence the thoughts and behaviour of others. Leadership is the process of influencing the thoughts and behaviour of others towards the goal setting and goal achievement.

FUNCTIONS OF LEADERS

There is no unanimity of opinion as to what the functions of leadership are. Generally speaking leadership functions are related to goal achievement and to the maintenance and strengthening of the group.

According to Barnard, a leader performs four main functions.

- Determination of objective
- Manipulation of means
- Control of the instrumentality of action; and
- Stimulation of coordination action

According to Dahama and Bhatnagar (1985)

- Executive
- Planner
- Policy maker
- Expert in human relations as well as technical field.
- External group representative
- Controller of internal relationship
- Purveyor of reward and punishment
- Arbitrator
- Exemplar
- Group symbol
- Surrogate of individual responsibility
- Idealist
- Father figure and
- Scape goat.

CLASSIFICATION OF LEADERS

Different authors classified the leaders in different ways. Some of the classifications are:

- **Beal, Bohlen and Raudabaugh (1962)** classified the leaders into four types as
  a. Born leaders
  b. Passive leaders (personal power or characteristic leader)
  c. Bureaucratic leaders and
  d. Democratic leaders

- **Mott (1972)** spoke about three major leadership categories
  a. Democratic
  b. Multifactor and
  c. Situational
- **Henning (1962)** analyzed the leadership and given the following three:
  a. Autocrat
  b. Bureaucrat and
  c. Neurocrat
- **Lester (1975)** also pointed three types as:
  a. Autocratic
  b. Democratic and
  c. Free rein
- **Haiman (1951)** described five categories as the
  a. Executive
  b. Judge
  c. Advocate
  d. Expert and
  e. Discussion leader
- **Carter (1961)** identified three leadership patterns according to the manner in which the leader was most oriented.
  a. Personal
  b. Institutional and
  c. Flexible
- **Sachdeva and Vidyabhushan (1974)** studied three main leadership types
  a. Authoritarian
  b. Democratic and
  c. Laissez-faire
- **Dahama and Bhatnagar (1985)** found several ways of classifying leaders, some of them are:
  a. Democratic, autocratic and laissez-faire
  b. Formal and informal leaders

c. Professional and lay (voluntary) leaders
d. Political, religious, social and academic
e. Elected, selected or nominated
f. Popular and unpopular
g. Traditional and progressive leaders

QUALITY OF LEADERS

Dahama and Bhatnagar (1985) brought out the determinants of effectiveness in leadership functions **(ten 'A's)**.

- Awareness of the functions of the group.
- Ability (self-adjustment) in performing the functions
- Achievement of goals
- Assignment of group functions to members
- Appraisal of effects of distribution of functions
- Accomplishment of different purposes under different environments
- Attainment of positive value as a leader
- Attaching a high code of conduct, ethical values and high morale in the group
- Arrangement of communication structures
- Acceptance of failure

Reddy (1987) tested the following traits or qualities as desirable for effective leadership.

- Physical fitness
- Mental ability (intelligence)
- Sense of purpose (having definite ideas regarding the aims of the group)
- Social insight (sensitivity to other person's position problems or points of view)
- Communication (including good listening and speaking acceptably in public)
- Love for people (friendliness without favoritism or without giving scope for indiscipline)

- Democracy (giving all members equal opportunities for participation, etc.)
- Initiative
- Enthusiasm
- Authority (based upon mastery of knowledge and skills in a particular field)
- Decisiveness (ability to make good and prompt decisions or judgments)
- Integrity or character
- Teaching ability
- Convictions and faith

ROLE OF LEADERS

Reddy (1987) brought out the following eight important roles.

- **Group Spokesman:** Leader has the responsibility to speak for the group and representing the group's interests and position faithfully and accurately.
- **Group Harmonizer:** All groups will usually have both uniformities and differences of opinion. The leader is responsible for pointing out to the group when potential conflict situations arise, that the common purpose is sufficiently worthy of co-operation that the differences be resolved peacefully.
- **Group Planner:** Generally persons are chosen for leadership positions because it is assumed that they know a little more about the problems confronting the group and their possible solutions than do the other members of the group. The group expects its leader to have new ideas for initiating activities. To meet this expectation leader must be able to plan, to visualize in his imagination the ways by which the group can satisfy its needs.
- **Group Executive:** Most groups have some established methods of conducting business and achieving consensus of opinion on issues that come up before them. The leader is the one who presides when the group is conducting business. As a group executive the leader is responsible for seeing that

the business of the organisation is carried on according to democratic principles.

- **Group Educator or Teacher:** Leader must share with the followers his knowledge and experience. Such sharing of experience and insight is teaching. Good leadership depends a large part upon teaching because the good teacher is not a dictator.
- **Symbol of Group Ideal:** All social groups have implicit or explicit norms or ideas. As a rule, persons accepted as leaders are those who have accepted these norms and live by them. The group expects his leadership to embody the ideals of group.
- **Group Discussion Chairman:** This role is more related to that of group executive. In recent years there has been an increased interest in group discussions. Having a panel, forum, conference etc. are common. In view of the increased use of these discussions this rule has been described.
- **Group Supervisor:** The leader has to work with followers and also with group organisations like youth clubs, co-operatives, farmer's associations, etc. Therefore this role becomes important for the leader.

Dahama and Bhatnagar (1985) ascertained the role expectations of farm leaders in view of the new strategy of agricultural production from the experts in the field of agricultural extension. By working out the index following were identified as the expected role of farm leaders.

- Raising demonstration plots and showing it to others
- Keeping themselves abreast with the latest agricultural technology
- Adopting themselves all the improved agricultural practices
- Organising meetings and other discussion groups
- Serving as marketing advisers
- Supplying the improved seeds to the farmers
- Assisting the farmers in getting credits
- Serving the innovative farmers in the village
- Educating the farmers to grow more than two crops a year

- Acting as liaison officers between the extension workers and the farmers.
- Educating the farmers in the latest agricultural technology.
- Helping the fellow-farmers in getting the needed agricultural inputs.

LEADERSHIP STYLE

While it is easy to understand what leadership is, it has been difficult to understand how successful leadership is to be exercised. In an effort to improve leader behaviour, and train in effective leadership, a large number of studies have been conducted. These may be grouped into three categories. 1. The early approaches to leadership. 2. Behavioural theories of leadership 3. Situational theories of leadership.

The early Approaches to Leadership

These theories explain leadership effectiveness in terms of the leader. They attempted to study the person of the leader and draw conclusions on leadership.

- **Great man approach:** This is the earliest theory of leadership, and suggested that great leaders were born and not made. They had some inborn qualities, and this made them great. However, there was no agreement of the qualities these leaders shared or how such people could be identified.
- **Trait approach:** In this approach, the researchers attempted to identify specific traits and characteristics of leaders. In more than a hundred studies from 1930 to 1950, attempts were made to correlate certain traits and the rise of leadership. A review of these studies grouped twenty-six traits identified in three or more studies. However, the application of the traits theory for development of leadership posed some problems. Often leaders (successful as well as unsuccessful!) and non-leaders possessed the same traits. This only shows that a given trait cannot cause an individual to become an effective leader. It is not possible to measure the extent of a trait a particular person has, and another situation. Although one can identify

certain traits good leaders appear to have, research indicates that no distinctive set of good leader traits exists.

Behavioural Theories of Leadership

These theories focus on the effectiveness of leader styles and behaviours.

- **Iowa studies:** In the University of Iowa, Lewin, Lippitt and White studied the impact of three leadership styles - autocratic, democratic, and laissez-faire. This relates to the behaviour of the leaders in relation to their group. The results indicated that 1) group members preferred democratic over autocratic leaders 2) incidents of intra-group hostility was higher in autocratic and laissez-faire groups. Incidentally, productivity was also higher in democratic group. This study was the turning point where research began to be focused on leadership behaviour rather than traits.
- **Ohio State studies:** The Ohio studies identified two dimensions of leader behaviour while directing the activities of the group towards achievement of the goal. These are initiating structure and consideration. Initiating structure refers to the leader's behaviour in delineating the relationship between himself and the members of the work group, and in establishing a well-defined pattern of organization, communication and methods of procedure. That is, the leader specified the task to be performed by each member of his group, set down deadlines, gave directions and put pressures for completion of the task. Consideration refers to behaviour of friendship, mutual trust, respect and warmth in the relationship between the leader and members of his group. The Ohio studies were the first to describe that the initiating structure and consideration were separate and distinct dimension. They were not a part of the same continuum, and can be plotted on two separate axis. A high score on one dimension does not necessitate a low score on the other. Any m ix of both dimensions could describe leader behaviour as illustrated in the following figure. This results in four leadership styles.

High consideration and low structure	High consideration and high structure
low structure and low consideration	high structure and low consideration
(Low) → Initiating structure → (High)	
The Ohio state Leadership Quadrant	

- **Michigan studies:** These studies attempted to compare the employee-centred and production-centred styles to determine which results in improved performance. No conclusive evidence was found to support either style. Higher productivity was observed in both the styles, by different studies. At first, the two styles were thought to be at the opposite ends of a continuums, however they were recognized as independent dimensions of leadership.
- **Managerial grid style:** Extending the Ohio and the Michigan studies, Blake and Mouton developed the concept of the managerial grid illustrated below:

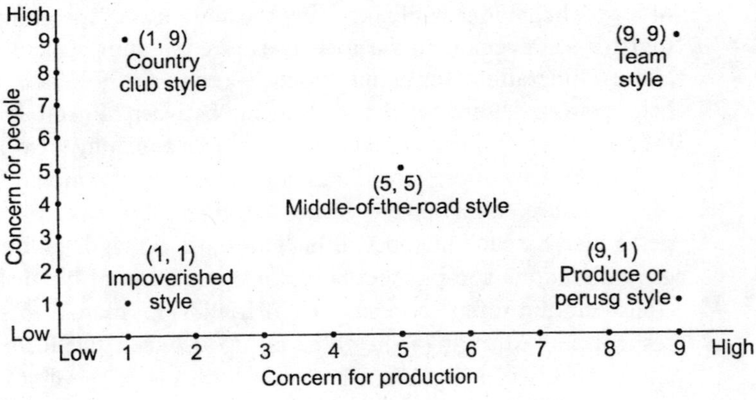

THE MANAGERIAL GRID LEADERSHIP STYLES

The five leadership styles illustrated are described below:

- **Impoverished (1-1):** Such managers have little concern for either people or production. Their involvement in jobs is the minimum. They exert only that little which is necessary to retain their positions. Their main function is to communicate information from superiors to subordinates and vice versa. They may be just marking time to retire or quit.

- **Country club (1-9):** This type of managers has little concern for production. Their only concern is for people. They turn thoughtful attention to satisfy their needs so that a happy and friendly organizational atmosphere and good work tempo are created and maintained.
- **Task (9-1):** These managers sometimes referred to as autocratic taskmasters, have little or no concern for people. Their only concern is with production, for which they arrange the conditions of work in such a way that human elements are kept at the minimum, and productive operations attain maximum efficiency.
- **Middle of the road (5-5):** These managers occupy a via media position. Their concern for both production and people is moderate. They strike a balance between the two. They neither set targets of production too high nor aim at building outstanding employee morale. The levels of both production and morale are on the whole satisfactory. The approach of these managers is generally one of benevolent autocracy.
- **Team (9-9):** These managers have the highest possible concern for both production and people. They gear the system of work to maximum efficiency and also inspire the subordinates to do the utmost. Individually they are committed to the work and together they feel a common stake in the organization. They develop a relationship of trust and respect. Team style is the best possible style, which integrates the needs of the individual, and those of the organization.

SITUATIONAL THEORIES OF LEADERSHIP

Studies concentrating on identification of qualities or behaviour of effective leaders could not completely understand the process of successful leadership that could be applied to various situations by different people. In fact, leadership effectiveness was now increasingly thought to be highly dependent on situations the leaders found themselves in. Thus new studies began to consider the influence of the leader, the followers, and the situation in an attempt to better understand the true nature of leadership.

- **Contingency theory:** The contingency model of leadership effectiveness developed by Fiedler postulated that leadership effectiveness is determined by the interaction of employee orientation (leadership style – relationship oriented or task oriented) with the three situational variables – leader member relations, task structure and leader position power. Leader - member relations are determined by the manner in which the leader is accepted by the group. Task structure reflects the clarity or ambiguity of the task. Clear task is said to have high structure. Position power refers to the authority vested in the leader because of his position in the organization.
- **Path-goal theory:** This theory focuses on the leader behaviour necessary to motivate to work more effectively. It derives from the expectancy theory, which assumes that motivation is a result of valence and expectancy. It becomes the leaders job to build both valence (amount of desire for a goal) and expectancy (strength of belief that particular act will be followed by a particular outcome) for the employee. Leaders encourage development of valence and expectancy by structuring the path towards goal attainment so that employees will see that their acts lead to their goals. Leaders can increase the rewards to subordinates, and also make the path easier by training and direction and by removing the barriers that stand in the way of goal attainment. Four leadership styles are used based on the contingent subordinate characteristics and environmental factors. The leadership styles are directive, supportive, participative and achievement-oriented.
- **Situational leadership theory:** This theory formulated by Hersey and Blanchard also identifies two dimensions of the leadership as in the Ohio State studies. These are 'relationship behaviour' and 'task behaviour'. However, unlike the managerial grid, which was also based on the Ohio State studies; this theory does not prescribe one best leadership style. The model states that the effectiveness of leader behaviour is situational. A manager should choose the style most appropriate for each situation. The appropriateness of the situation is judged by the maturity of the followers.

The model is illustrated below: **The four leadership styles** are described below:

- **Telling (S1):** Provide specific instruction to subordinates and closely supervise performance.
- **Selling (S2):** Explain decision to subordinates and provide opportunity for clarification
- **Participating (S3):** Share ideas with subordinates and facilitate the decision-making process.
- **Delegating (S4):** Turn over responsibility for decisions and implementation to subordinates.

LEADERSHIP STYLE SUITABLE TO INDIAN CONDITIONS

Singh (1982) studies the profile of leadership styles in India and found that Indian managers do not have a dominant preference for a particular style and suffered from a style flux and normlessness. Leadership is viewed in terms of functions and roles the managers are supposed to perform by the managerial community overlooking the two important elements, the leadership dominance and consistency. The ambiguity about the concept of leadership style and the lack of proper emphasis on leadership dominance and consistency may be responsible for the style flux according to Singh (1982). Style normlessness can be explained in terms of too much emphasis on situational factors. Leadership and supervision are not the same. Leadership involves guiding, helping, planning and developing people from one phase of growth to a higher one while supervision entails policing i.e. keeping close watch on the organizational activities. In India some of the researchers have recommended authoritarian style linking leadership to the profile of working class and with the assumptions that the workers are dependence-prone, cannot take decisions on their own, need to be goaded, controlled and guided which are not based on realities. The works of Singh and Dass (1977) and Sharma (1971, 1976) clearly demonstrate that the working class also desire autonomy, freedom, responsibility and challenges for motivation. The effectiveness of the leadership style is judged against the leader's ability to influence others' action and behaviour, which depends upon his position on the power continuum.

The place and position on the power continuum helps one to prescribe a functional and effective style suitable for Indian conditions.

Basically the following **six factors** determine a leader's position on the power continuum:

- **Leaders own strengths, weaknesses and charisma:** Family background, educational background, technical expertise, nearness to power centre and charismatic personality contribute to the personal strengths. Few leaders have the charisma capable of evoking reverence and the ability of a leader to operate on charisma becomes high if the profile of subordinates is that of non-questioning. Historically, the caste system in India where authority was blindly accepted was contributory to charismatic authority. This is fast disappearing due to political, social and economic policies followed by the Government.
- **Profile of subordinates:** The relevance of the subordinate's profile in determining the managerial style can be viewed mainly from two angles: (1) Subordinates' respect for the superior's authority and decisional judgment (2) The subordinate's motivational profile. The respect for superior's authority has, of late, eroded considerably, both in the industrial and social institutions due to the credo for permissive values and the stronger desire for participative decision-making. Emergence of trade unions, associations and government's regulatory measures, etc., has led to a virtual revolt against autocratic and authoritarian managerial style. A study of the motivational profile at top, middle and lower management levels by Singh (1982) have shown the preponderance of freedom, autonomy, responsibility, and challenges and varieties over other motivational factors at all the levels of management.
- **Socio-political environment:** Political democracy has taken deep roots in Indian life and for such environment the best-suited leadership style would be participative and democratic.
- **Organizational power position** If the power to punish or reward, and if autonomy and operational flexibility given to a manager by the organization is high, it helps him to get overtly

an unquestioned obedience from his subordinates. Now a days this power of the managers has been considerably reduced due to trade union activities and governmental controls. In such situations the managers (leaders) have to be more democratic and rely less on organizational authority.

- **Knowledge and expert power:** Managers with superior job knowledge or expertise are capable of exerting greater influence on their subordinates. Today the knowledge differentiation between the leaders and followers is getting fast reduced and in fact, an opposite trend can be seen i.e. the younger generation of subordinates has better exposure and academic qualifications. The young subordinates are more informative and better trained. The knowledge obsolescence is more common among the senior managers. The younger generations of today have a strong desire to question the unilateral decision making process of the superiors, thereby putting enormous limitations on autocratic style i.e. centralized decision-making system.
- **Nature of task:** If the job technology is highly programmed and structured making performance qualification easier, the superior exercises greater control over his subordinates. In case of unstructured and interdependent job technology, work quantification is difficult which greatly reduces the control of manager over the behaviour of his subordinates.

LEADERSHIP STYLE BASED ON AUTHORITY (3 TYPES)

Autocratic

Characteristics

- Determines all policies, activities and goals of the organisation.
- Takes no part in work except when conducting meeting, telling others what to do or demonstrating.
- Members are uncertain about what to do and usually take to take.
- Leader is personal in both praise and criticism of the work.

Democratic

Characteristics

- Produces a shared leadership that permits a feeling of satisfaction and achievement.
- Helps the members of understand the steps required in working.
- Members take more responsibility for group maintenance and task performance.

Laissez-faires or Free-rein

Characteristics

- Gives minimum guidance
- Remains in the background and seldom express an opinion or works with a minimum of rules.
- Members often act as leaders in making decisions that guide the organisation.

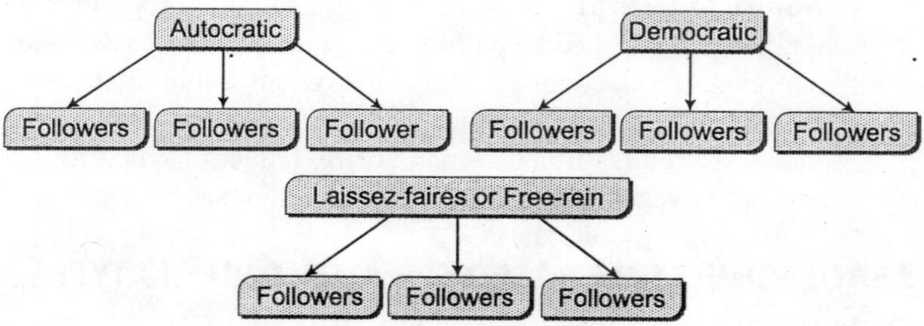

ADVANTAGES OF LEADERS IN EXTENSION

Reddy (1987) brought out the following advantages and limitations in using local leaders in extension.

Advantages

- Local leaders virtually play the role of extension teachers and hence the volume of coverage with improved practice is increased.

- Cost of extension is reduced, as local leaders are not paid for their work.
- Local leaders themselves become better taught, because of the experience they gain in teaching and influencing others.
- People accept a new idea more readily from a local person who has practically tried it, while they may resist if the ideas were to come from a paid extension workers.
- The frequent association of extension personnel with the local leaders enhances his prestige.

Limitations

- Persons selected as leader may not have the expected following among neighbors.
- Sometimes they may not be willing to devote required time to work or may be a poor teacher.
- Considerable time is required to locate and train local leaders.
- Local leaders may try to use prestige with position for personal advantage.
- Public recognition and publicity given to informal local leaders may sometimes jeopardies their position and adversely affect their influence among the people.

CHAPTER 8

GROUP DYNAMIC AND TEAM BUILDING

INTRODUCTION

People may underestimate the importance of society and group memberships on their lives. Whilst people sometimes undertake solo journeys yet by and large much of our experiences of life involves being engaged with others and groups. Within an organization we do find number of groups. Individuals joining group (s) is a reality – may be formal or informal groups. People work in groups quite frequently and in many different areas of their life e.g. at work, school/college, sport, hobbies. The managers need to understand group dynamics that can enable managers to adopt the right approach of interacting with them.

Group Dynamics

- Group dynamics deals with the attitudes and behavioral patterns of a group. Group dynamics concern how groups are formed, what is their structure and which processes are followed in their functioning. Thus, it is concerned with the interactions and forces operating between groups.
- Is the study of interactions and forces within small face-to-face groups. This interaction may help or hinders the function of the group.
- The social process by which people interact face-to-face in groups is called group dynamics.
- A group consists of two or more people who have:
  a. A common objective or task
  b. An awareness of group identity and boundary
  c. A minimum set of agreed values and norms

- Group dynamic is a system of behaviors and psychological processes occurring within a social group (*intra* group dynamics), or between social groups (*inter* group dynamics).

History of Group Dynamics

- In **1924, Gestalt psychologist Max Wertheimer** identified this fact, stating 'There are entities where the behaviour of the whole cannot be derived from its individual elements nor from the way these elements fit together; rather the opposite is true: the properties of any of the parts are determined by the intrinsic structural laws of the whole' (Wertheimer 1924, p. 7) As a field of study, group dynamics has roots in both psychology and sociology.
- **Wilhelm Wundt (1832–1920),** credited as the founder of experimental psychology, had a particular interest in the psychology of communities, which he believed possessed phenomena (human language, customs, and religion) that could not be described through a study of the individual.
- **Emile Durkheim (1858–1917),** who was influenced by Wundt, also recognized collective phenomena, such as public knowledge.
- **Gustave Le Bon (1841–1931)** who believed that crowds possessed a 'racial unconscious' with primitive, aggressive, and antisocial instincts, and William McDougall (psychologist), who believed in a 'group mind,' which had a distinct existence born from the interaction of individuals.
- **Kurt Lewin (1890–1947**) who coined the term group dynamics to describe the positive and negative forces within groups of people.
- In 1945, he established **The Group Dynamics Research Center** at the Massachusetts Institute of Technology, the first institute devoted explicitly to the study of group dynamic. Throughout his career, Lewin was focused on how the study of group dynamics could be applied to real-world, social issues. An increasing amount of research has applied evolutionary psychology principles to group dynamics. Humans are

argued to have evolved in an increasingly complicated social environment and to have many adaptations concerned with group dynamics. Examples include mechanisms for dealing with status, reciprocity, identifying cheaters, ostracism, altruism, group decision, leadership, and intergroup relations.

Key Theorists of Group Dynamics

- **Gustave Le Bon:** Gustave Le Bon was a French social psychologist whose seminal study, *The Crowd: A Study of the Popular Mind* (1896) led to the development of group psychology
- **William McDougall:** The British psychologist William McDougall in his work *The Group Mind* (1920) researched the dynamics of groups of various sizes and degrees of organization.
- **Sigmund Freud:** In *Group Psychology and the Analysis of the Ego,* (1922), Sigmund Freud based his preliminary description of group psychology on Le Bon's work, but went on to develop his own, original theory, related to what he had begun to elaborate in *Totem and Taboo.* Theodor Adorno reprised Freud's essay in 1951 with his *Freudian Theory and the Pattern of Fascist Propaganda*, and said that "It is not an overstatement if we say that Freud, though he was hardly interested in the political phase of the problem, clearly foresaw the rise and nature of fascist mass movements in purely psychological categories.
- **Jacob L. Moreno:** Jacob L. Moreno was a psychiatrist, dramatist, philosopher and theoretician who coined the term "group psychotherapy" in the early 1930s and was highly influential at the time.
- **Kurt Lewin:** Kurt Lewin (1943, 1948, 1951) is commonly identified as the founder of the movement to study groups scientifically. He coined the term group dynamics to describe the way groups and individuals act and react to changing circumstances.

- **William Schutz:** William Schutz (1958, 1966) looked at interpersonal relations as stage-developmental, inclusion (am I included?), control (who is top dog here?), and affection (do I belong here?). Schutz sees groups resolving each issue in turn in order to be able to progress to the next stage.
- **Wilfred Bion:** Wilfred Bion (1961) studied group dynamics from a psychoanalytic perspective, and stated that he was much influenced by Wilfred Trotter for whom he worked at University College Hospital London, as did another key figure in the Psychoanalytic movement, Ernest Jones. He discovered several mass group processes which involved the group as a whole adopting an orientation which, in his opinion, interfered with the ability of a group to accomplish the work it was nominally engaged in. His experiences are reported in his published books, especially *Experiences in Groups.* The Tavistock Institute has further developed and applied the theory and practices developed by Bion.
- **Bruce Tuckman:** Bruce Tuckman (1965) proposed the four-stage model called Tuckman's Stages for a group. Tuckman's model states that the ideal group decision-making process should occur in four stages:
  a. Forming
  b. Storming
  c. Norming
  d. Performing
  e. Adjourning
- **M. Scott Peck:** M. Scott Peck developed stages for larger-scale groups (i.e., communities) which are similar to Tuckman's stages of group development. Peck describes the stages of a community as:
  i. Pseudo-community
  ii. Chaos
  iii. Emptiness
  iv. True Community
- **Richard Hackman:** Richard Hackman developed a synthetic, research-based model for designing and managing work

groups. Hackman suggested that groups are successful when they satisfy internal and external clients, develop capabilities to perform in the future, and when members find meaning and satisfaction in the group. Hackman proposed five conditions that increase the chance that groups will be successful.

Intragroup Dynamics

- Intragroup dynamics (also referred to as in-group-, within-group, or commonly just 'group dynamics') are the underlying processes that give rise to a set of norms, roles, relations, and common goals that characterize a particular social group.
- Examples of groups include religious, political, military, and environmental groups, sports teams, work groups, and therapy groups. Amongst the members of a group, there is a state of interdependence, through which the behaviours, attitudes, opinions, and experiences of each member are collectively influenced by the other group members.

Intergroup Dynamics

- Intergroup dynamics refers to the behavioural and psychological relationship between two or more groups. This includes perceptions, attitudes, opinions, and behaviours towards one's own group, as well as those towards another group.
- In some cases, intergroup dynamics is prosocial, positive, and beneficial (for example, when multiple research teams work together to accomplish a task or goal).
- In other cases, intergroup dynamics can create conflict.

Groups is a unit of two or more individuals who are in reciprocal communication and interaction with other. A social group is a collection of individuals two or more interacting with each other who have common set of objectives participate in similar activities

Social Group Definition According to Different Authors

- **Bogardus:** It is a number of units anything in close proximity to one another.

- **Mac Iver:** Any collection of human beings who are brought into social relationships with one another.
- **Sheriff and Sheriff:** A group is a social unit which consists of number of individuals who stand in, more or less, definite status and role relationships to one another and which possess a set of values or norms of its own, regulating the behaviour of individual members at least in matters of consequence to the group.
- **Williams:** A social group is a given aggregate of people, playing inter-related roles and recognized by themselves or others as a unit of interaction.

Characteristics of Social Group

- **Relationship:** Members of group are inter-related to each other. Reciprocal relations form an essential feature of a group.
- **Sense of unity:** The members of the group are united by a sense of unity and a feeling of sympathy.
- **We-feeling:** The members of a group help each other and defend their interests collectively.
- **Common interest:** The interests and ideas of the group are common. It is for the realization of common interests that they meet together.
- **Similar behaviour:** The members of a group behave in a similar way for the pursuit of common interest.
- **Group norms:** Every group has its own rules or norms which the members are supposed to follow.

CLASSIFICATION OF GROUP

Social groups have been classified in various ways by different authors.

- **According to Dwight Samderson**
    a. **Involuntary group** is based on kinship such as family. A man has no choice to what family he will belong.
    b. **Voluntary group** is one which a man joins of his own volition. He agrees to be a member of it and is free to withdraw at any time from its membership.

c. **Delegate group** is one which a man joins as a representative of a number of people either elected by them or nominated by some power. Parliament is a delegate group.

- **According to Tonnies:** He classified the groups into communities and associations.
- **According to Cooley** (C. H. Cooley classified the groups)

  a. **Primary group:** Face to face and intimate relationship such as in a family.

  b. **Secondary group:** A political party the relationships are indirect, secondary or impersonal.
- **According to F. H. Giddings**

  a. **Genetic group:** A family in which a man is born involuntarily.

  b. **Congregate group:** The voluntary group into which one moves or joins voluntarily.
- **According to George Hasen**

  a. **Unsocial group** is one which largely lives to itself and for itself and does not participate in the larger society of which it is a part. It does not mix up with other groups and remains aloof from them.

  b. **Pseudo-social group** participates in the larger social group but mainly for its gain and not for the greater good.

  c. **Anti-social group** is one that acts against the interest of society. A group of students that destroys public property is an anti-social group. Similarly a political party that plan to over throw a popular government is anti-social.

  d. **Pro-social group** is the reverse of anti-social. It works for the larger interests of the society. It is engaged in constructive tasks and concerned with increasing the welfare of all the people.
- **According to Sorokin**

  a. **Horizontal groups:** Large, inclusive groups such as nations, religions, organization, political parties etc.

  b. **Vertical groups:** Small divisions such as economic classes. Since the vertical group is a part of the horizontal group (the larger group) an individual is member of both

- **According to Charles A. Ellwood**
  a. Involuntary and voluntary groups
  b. Institutional and non-institutional groups
  c. Temporary and permanent group
- **According to Leopold**
  He classified groups into:
  a. Crowd
  b. Groups and
  c. Abstract collectives.
- **According to Park and Burgers**
  They distinguished the groups as territorial and non-territorial groups.
- **Lowis Gillin and Philip Gillin (Gillina and Gillin)**
  a. Blood relationship
  b. Bodily characteristics
  c. Physical proximity and
  d. Culturally derived interests
- **According to summer**
  a. In group is the group with which the individual identifies himself, his family, tribe, sex, college, occupation, religion, etc. by virtue of his awareness of likeness. It has inclusion of some persons and exclusion of other persons. It has the collective pronoun 'we'. It has some sympathy and a sense of attachment to other members of the group.
  b. Out group is defined in contrast between 'they' and 'we', like we are democrats and they are communists. We are Hindus and they are Muslims. We are Brahmins and they are Nadars. Such attitudes that "these are my people" and that "those are not my people", produce a sense of attachment to other members in group while a sense of indifference and even antagonism to the members of out group.

TEAMS

A team is a group of two or more individual which is very cohesive, works unitedly, and its members perfectly understand each other's contribution and their own contribution for successful completion on the team's goals.

Building a team and sustaining it through different projects and business tasks can be both a rewarding and frustrating experience for everyone concerned. Adding to this frustration is the fact that most businesses are never shown how they can build a successful team, often developing one through trial and error. Although this might achieve the desired result, it can cost the business, and team members, much time, energy and financial resources trying to get the team mix right.

Team Work

Teamwork is a collective activity of a group of two or more individuals who are totally committed to the team's goals, and who collaborate with each other resulting in coordinated and united efforts, which is essential for the successful completion of a given task.

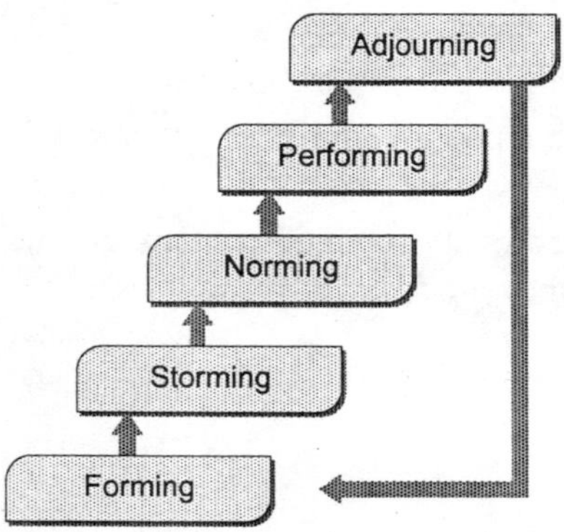

The three most important characteristics of good teamwork are:

- **Unity of direction:** Resulting thorough understanding of the total task, one's own role and others role in it.
- **Cooperation:** Resulting from common point of view about, and commitment to the total task.

- **Subordination of personal prominence:** Resulting from commitment to the success of the total task.

Stages of Team Formation (According to Bruce Tuckman, 1965)

Forming (Orientation): Simply the bringing together of a group of individuals.

Characteristics

- Excitement, anticipation, and optimism.
- Pride in being chosen for the project.
- A tentative attachment to the team.
- Suspicion and anxiety about the job.
- Defining the tasks and how they will be accomplished.
- Determining acceptable group behaviour.
- Deciding what information needs to be gathered.
- Abstract discussions of the concepts and issues, and for some members, impatience with these discussions.
- There will be difficulty in identifying some of the relevant problems.
- Moderately eager.
- Have generally positive expectations.
- Have some anxiety about why they are there and what it all means.
- Have some anxiety about other members such as who they are and what they are like.

Storming (Dissatisfaction)

Characteristics

- Argument
- Conflict
- A dip in morale
- Resisting the tasks
- Resisting quality improvement approaches suggested by other members

- Sharp fluctuations in attitude about the team and the project's chance of success
- Arguing among members even when they agree on the real issues
- Defensiveness, competition, and choosing sides
- Questioning the wisdom of those who selected this project an appointed the other members of the team
- Establishing unrealistic goals
- Disunity, increased tension, and jealousy

Norming (Resolution)

Stage in the group's development involves.

Characteristics

- An ability to express criticism constructively.
- Acceptance of membership in the team.
- An attempt to achieve harmony by avoiding conflict.
- More friendliness, confiding in each other, and sharing of personal problems.
- A sense of team cohesion, spirit, and goals.
- Establishing and maintaining team ground rules and boundaries.
- Resolving of issues
- Setting up group processes
- Setting of group policies, procedures, and values
- Increasing production

Performing (Production)

Characteristics

- Members have insights into personal and group processes, and better understanding of each other's strengths and weakness
- Constructive self-change
- Ability to prevent or work through group problems
- Close attachment to the team

- The team is accomplishing work effectively.
- Production is high and the climate is positive.

Adjourning (Termination)

Characteristics

- In the case of temporary teams such as task forces, design teams, and problem solving teams, a fifth stage reflects the ending of the process.
- Depending on the team's success in accomplishing its task and how strongly the members have bonded, this stage may reflect either a sense of loss or relief. When a team ends, time should be spent addressing how it should be done to properly recognize the team's accomplishments.

Desirable Climate for Teamwork

A desirable climate for teamwork will have the following requirements:

- The atmosphere in which the teamwork is relaxed, comfortable and informal.
- The people involved can easily laugh and joke together and there is a lot of backslapping and bonhomie.
- The task is well understood by all the members, and their role in its is accepted by one and all.
- The members communicate a lot and listen well to each other
- A lot of task relevant discussion takes place.
- In a tam, the members express both their ideas and their feelings.
- Conflicts and disagreements center around ideas and methods, and not around people.
- Decisions are usually based on consensus.
- When actions are decided upon, clear assignments are made and accepted by the members.

8 Ways or 8 Practices to Build a Great Team (Lynda and Tamara 2007)

- **Executive support:** Teams do well when executives in the organization invest in supporting social relationship, demonstrate collaboration behavior themselves and create gift culture; a culture in which employee interact with top executives and colleagues; a good interaction gives value (gift).

- **Investing in signature relationship practices:** Signature Practices–that were memorable, different for others to replicate and particularly well suited to their own business environment creating a gift culture – for executives to ensure that mentoring and coaching become embedded in their own routine behavior and throughout the company.
- **Focused HR Practices:** Ensuring the requisite skills: It is not only collaboration required to success of a team relative skills are necessary to be more collaborative and to know how to work well in a team. Crucial skills – appreciating others, being able to engage in purposed conservation, productivity and creativity resolving conflicts and program management E.g. PWC.
- **Supporting a sense of community:** Sponsoring events, family gathering to build a good network and community within group and team. The right team leaders – where collaboration level is high, team leader plays a significant role.
- **Assigning leaders who are both task and relationship oriented:** It is mentioned that most of successful teams were led by people who were task and relationship oriented both. In earlier stage they were task oriented, when project was on development stage they change their style into relationship oriented roles.
- **Team formation and structure:** Building on heritage relationship: Building a heritage relationship within team enhances the chances of a project's success. New members who are new at the time of formation find more difficulties to create high collaborative team than those with established relationship. E.g. Nokia For a new project, rather than shifting some team members, shifting of whole team. If heritage relationship is not managed skillfully too many of them can actually disrupt collaboration.

- **Understanding role clarity and task ambiguity:** Collaboration improves when role of individuals are clearly defined and well understood, without such clarity, team members waste their energy and time in negotiating their roles.
- **Organizational culture or business environment:** Models for teams need to Models for teams need to be realigned with the demands of the current business environment.

Causes of Poor Teamwork or the Lack of Teamwork

- Lack of understanding of the total task by the members of the team.
- Lack of concern for others problems.
- Lack of proper plan for the completion of the task.
- Lack of trust and openness between the members of the team.
- Lack of confidence by some of the team members in the capability of others.
- Fear in the minds of the members that others may misuse the resources.
- A high degree of possessiveness by some of the members of the team.
- The bloated ego of some of the members which causes problems.
- Misunderstandings arises between the members of the team.
- People fearing the loss of control over resources when shared with others.
- Genuine difficulty and inconvenience faced by people.
- Operational problems in catering to the needs of all.
- Inadequacy of the resources available in the organization.
- Poor leadership, which is unable to keep the team together.

Five Dynamics of Teamwork and Collaboration

- Team member
- Team relationships

- Team problem solving
- Team leadership
- Organizational environment

Five Steps to Effective Problem Solving

- Identify the problem.
- Create a collaborative setting:
  a. Surface any assumptions and biases
  b. Agree on principles for discussion
- Identify and analyze the issues.
- Identify possible solutions.
- Resolve the single question.

CHAPTER 9

INTERPERSONAL COMMUNICATION AND RELATIONSHIP

COMMUNICATION

Communication is the process of sharing our ideas, thoughts, and feelings with other people and having those ideas, thoughts, and feelings understood by the people we are talking with. When we communicate we speak, listen, and observe. Communication is the exchange of thoughts, messages, or information as by speech, signals, writing or behavior. Derived from the Latin word "communis", meaning to share. Communication requires a sender, a message. And a recipient, although the receiver need not be present or aware of the senders intent to communicate at the time of communication; thus communication can occurs across vast distances in time and space. Communication requires that the communicating parties share an area of communicative commonality. The communication process is complete between participants.

The way we communicate is a learned style. As children we learn from watching our parents and other adults communicate. As an adult we can learn to improve the way we communicate by observing others who communicate effectively, learning new skills, and practicing those skills. The ability to effectively communicate at work, home, and in life is probably one of the most important sets of skills a person needs. What would our life and world be like without communication? We cannot get along without it. It is also not easy, and we all have probably had experiences where our communication failed or ran into a barrier. So, if we can understand the communication process better and improve it, we will become a more effective and successful communicator.

COMMUNICATION PROCESS

This lesson teaches learners to define the elements of effective communication and its process. It will focus on communication as the best way to convey meaning and introduce barriers to the communication process. Students will learn the communication process loop and its key elements. Now that you understand the purposes of effective communication, let's take a closer look at the elements in the communication process. By analyzing the parts of the process, we are better able to understand the whole. The communication process is a simple model that demonstrates all the factors that can affect communication. Communication is effective if the message that is received is the same one that is sent.

Communication Process Key Elements

- **Sender:** The communicator or sender is the person who is sending the message. There are two factors that will determine how effective the communicator will be. The first factor is the communicator's attitude. It must be positive. The second factor is the communicator's selection of meaningful symbols, or selecting the right symbols depending on your audience and the right environment. Talk about a few wrong examples.
- **Message:** A communication in writing, in speech, or by signals.
- **Receiver:** The receiver is simply the person receiving the message, making sense of it, or understanding and translating it into meaning. Now think about this for a moment: the receiver is also a communicator. How can that be? (When receiver responds, he is then the communicator.) Communication is only successful when the reaction of the receiver is that which the communicator intended. Effective communication takes place with shared meaning and understanding.
- **Feedback:** Feedback is that reaction I just mentioned. It can be a verbal or nonverbal reaction or response. It can be external feedback (something we see) or internal feedback (something we can't see), like self-examination. It's the feedback that allows the communicator to adjust his message and be more effective. Without feedback, there would be no way of

knowing if meaning had been shared or if understanding had taken place.

Lets discuss communication as a two way process. The information goes out to a person on the other end. There is a sender and a receiver. Simply put, effective communication is getting your message across to the receiver. It is the sender's responsibility to make sure that the receiver gets the message and that the message received is the one sent. Communicating is not an isolated series of one skill, it involves several skills. For example, speaking involves not only getting your message across but also being able to listen and understand what others are saying (active listening) and observing the verbal and nonverbal clues in order to monitor the effectiveness of your message. Communication is a two-way process that involves getting your message across and understanding what others have to say. Communication involves active listening, speaking and observing. Now that you have learned the communication process, you can begin to evaluate your communication skills. Begin to watch yourself in action. Each time you communicate observe what you do, how it went, what went well, and what could have been better.

Communication is the art of sharing ideas, information, instruction or feelings. The basic ingredients of good communication are clear thinking, clear speaking and clear writing. We spend about 75 per cent of our waking hours in some form of communication such as talking, listening, reading and writing.

- **Interpersonal communication** is the ongoing, ever-changing process that occurs when you interact with another person, forming a dyad, which is defined as two people communicating with each other.
- **Direct interpersonal communication** involves a direct face-to-face relationship between the sender and receiver of a message, who are in an interdependent relationship. Because of interpersonal communication's immediacy (it is taking place now) and primacy (it is taking place here), it is characterized by a strong feedback component. Communication is enhanced when the relationship exists over a long period of time. Interpersonal communication involves not only the words used but also the various elements of nonverbal communication.

The purposes of interpersonal communication are to influence, help and discover, as well as to share and play together.

Why Do We Need Interpersonal Communication?

- We cannot be human alone.
- We live in a world filled with other people.
- We live together, work together and play together.
- We need each other for security, comfort, friendship and love.
- We need each other to mature through dialogue.
- We need each other to achieve our goals and objectives.
- None of these needs could be addressed without interpersonal communication.

Interpersonal Communication can be Categorized by the Number of Participants

- **Dyadic communication involves two people.** Example: Two friends talking.
- **Group communication involves three or more persons**, though communication scholars are inconsistent as to the top end of the number scale. The smaller the number in the group, the more closely this mode resembles interpersonal communication. Often group communication is done for the purpose of problem solving or decision making. Example: University study group.
- **Public communication** involves a large group with a primarily one-way monologue style generating only minimal feedback. Information sharing, entertainment and persuasion are

Common purposes of public communication. Example: Lecture in university class.

Another way of categorizing interpersonal communication is on the function or setting of the communication.

- Organizational communication deals with communication within large organizations such as businesses. This is sometimes considered as a part of group communication, but communication scholars have built up a body of knowledge focused primarily on organizations. Example: Work focused discussion between employer and employee.

- Family communication focuses on communication patterns within nuclear, extended and blended families. Like organizational communication, this too is sometimes seen as part of the general category of group communication, but much research has been focused specifically on communication within a family relationship. Family communication can be enhanced by the long-standing and close relationships among participants as well as the likelihood that families have shared heritage, similar values, and social rituals. Patterns differ in communication between spouses, between parent and child, among siblings, and within the wider family context. Example: Conversation during a holiday meal.

Stages of Interpersonal Communication (Mark Knapp)

These stages can be applied to personal friendships, romantic relationships, business encounters, and many other types of interaction.

- **The initial encounter:** Offers a first impression that can be full of communication data. Likes or dislikes can be instantaneous, though many people have learned that first impressions may be misleading.
- **Experimenting:** The second step in interpersonal communication. In this step, information is exchanged on a variety of topics – general and "safe" at first, gradually becoming more personal and more self-revealing.
- **Intensifying:** When the experimentation leads to positive mutual conclusions that the relationship is worth pursuing.
- **Integration:** In which mutual decisions are made that the relationship is fulfilling. This is the stage of intense friendships, close business partnerships, romantic commitments, and so on.
- **Bonding:** Is the final stage, in which the relationship is sealed (often formally with contracts written agreements) and generally is publicized (such as through announcements)

The Essential Elements of Interpersonal Communication

- **People:** The senders and receivers of communication messages
- **Messages:** The content of communication
- **Channels:** The media through which messages travel
- **Noise:** Interference with the ability to send or receive messages
- **Feedback:** Information received in exchange for messages
- **Context:** The environmental,
- **Effect:** The result of a communication episode

Types of Noise

- **Semantic noise:** Noise due to the failure to understand the intended meaning of one or more words or the context in which the words are being used (persons speaking different languages, using jargon and "technicalese").
- **Physiological noise:** Noise due to personal illness, discomfort, or a physical problem including speech, visual, auditory, or memory impairment (difficulty articulating, hearing or sight loss, fatigue, disease).
- **Psychological noise:** Noise due to anxiety, confusion, bias, past experience, or emotional arousal that interferes with communication (sender or receiver prejudice, closed-mindedness, rage).
- **Intellectual noise:** Noise due to information overload or underload (over- or under preparedness).
- **Environmental noise:** Noise due to the sound, smell, sight, and feel of the environment or physical communication space that distracts attention from what is being said or done (cars honking, garbage rotting, people talking at once, cellular or computer interference).

Model of Interpersonal Communication

- **Linear model of interpersonal communication:** To be sure, the thinking about interpersonal communication has evolved over the years. The earliest mode - a linear or unidirectional model- depicted communication as going in one direction only. Questions such as "Did you get my message?," statements such as "I gave you that idea," and acts such as leaving someone a note with instructions give voice to this one-way perspective. While containing many of the elements identified earlier, notice that the linear model omits both feedback and context.

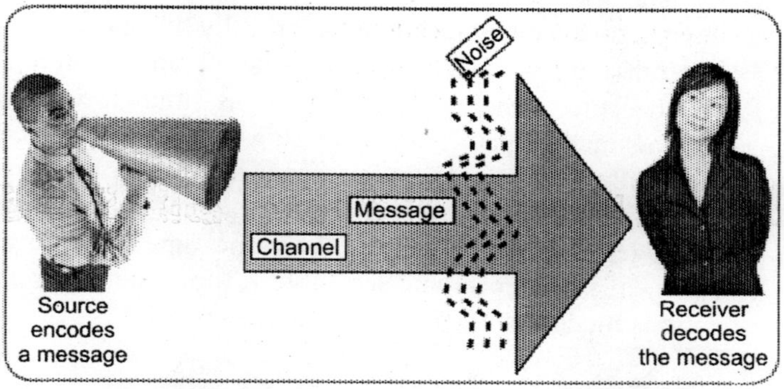

- **Interaction model of interpersonal communication**: Gradually, a more realistic two-way model - known as an interaction mode - came to be the model of choice. The interaction model visualizes interpersonal communication not as a one-way event but as a back-and-forth process, much like a game of tennis; it also acknowledges the presence and effects of both feedback and context. However, though more accurate than the one-way model, the interaction model fails to capture the complexity of interpersonal communication, including the reality that interpersonal communication does not involve just a back-and-forth action and reaction as might occur when you send a text and your friend responds.

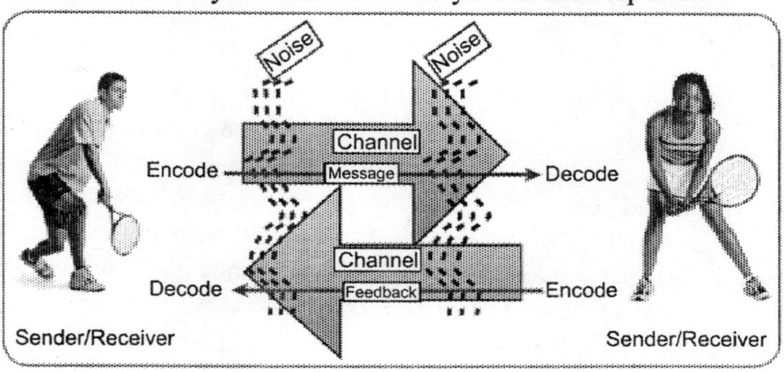

- **Transactional model of interpersonal communication:** Many communication exchanges involve source and receiver responding to one another simultaneously rather than sequentially. Thus, a more recently developed and even more realistic way to visualize how the elements at work during interpersonal communication dynamically relate to each other is a transactional model. The transactional model's strength is that it depicts sending and receiving as simultaneous rather than distinctly separate acts. By doing so, it helps us visualize the vital complexity of interpersonal interaction by showing us that source and receiver send messages to and receive messages from each other at the same time, reflecting the reality of a conversation. (See Table 1.3 for a summary of the various models' strengths and weaknesses.)

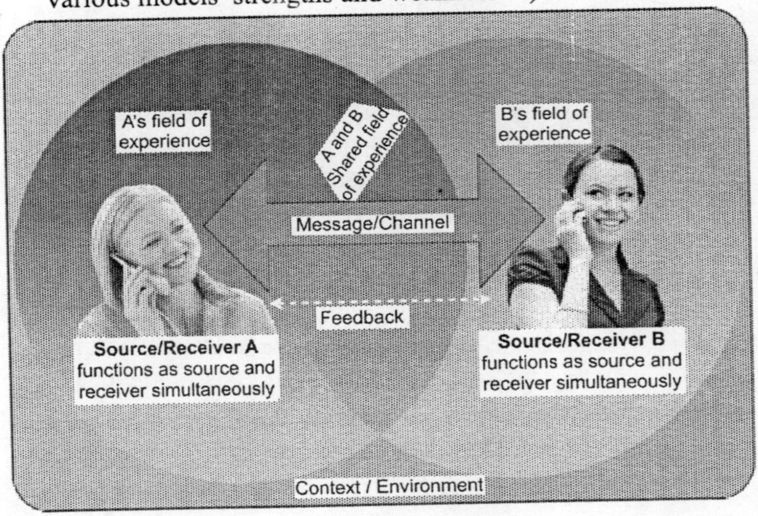

- **Two-person relationship model:** The view we have of ourselves and others will also affect the communications that take place. Turner and Lombard's two-person relationship model below highlights the various aspects at work. Imagine a situation where Person A is lacking in confidence in their job, and is intimidated by the experience and track record of person B. Likewise person B may be very sure of themselves and somewhat dismissive as Person A. (Source: Interpersonal

Behaviour and Administration, Arhtu N. Turner & George F. F. Lombard, Free Press 1969, New York) When sitting down to discuss a work issue, these relationships will colour the conversation. Even if person A puts forward the best solution to a particular problem, it probably won't get accepted because of the relationship issues involved.

Two-Person Relationship Model

Turner and Lombard

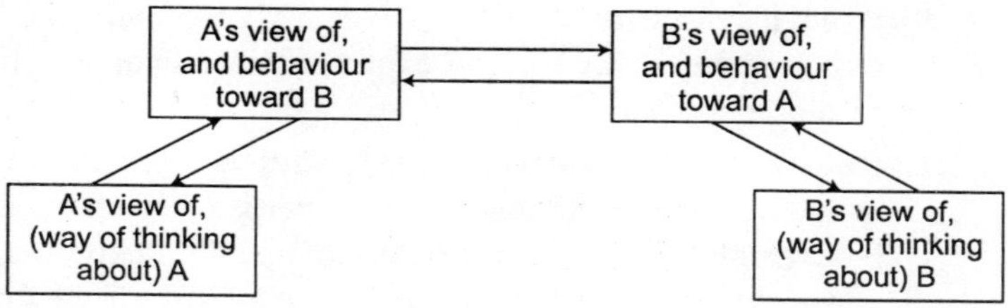

INTERPERSONAL STYLES

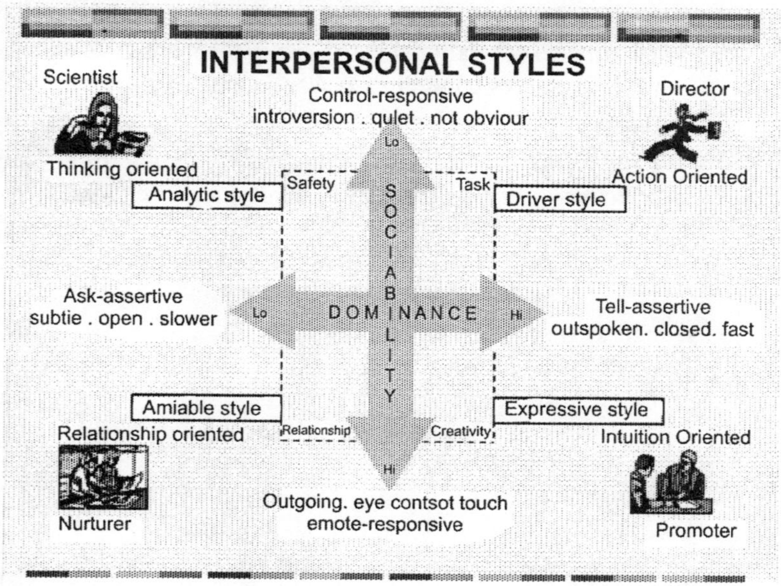

THE FOUR SOCIAL (INTERPERSONAL) STYLES

- **Analytic**
  - a. **Exacting, orderly, precise: Relies on facts and logic:** Explores all avenues before making a decision. Preferred leadership style based on competence, not force. Completes projects in a focused manner. In stressful situations may avoid personal involvement.
  - b. **Moderate, restrained, methodical:** Works best alone and on clearly defined tasks. Good listener. Tends to influence through reason rather than power. May be reluctant to direct others. May rely on expertise more than assertion to succeed.
  - c. **Reserved, conservative, orderly:** Prefers a situation that calls for independent analysis or expertise. Tends to lead by example. Enjoys a leadership role in which know-how and technical expertise are important. Will strive for a logical solution.
  - d. **Task-oriented, cooperative, practical:** Tend to depend on themselves to get the job done, but recognizes the importance of relationships. Usually responds to feedback well. Resolves conflict by reasoning and diplomacy.
- **Amiable**
  - a. **Unpretentious, friendly, consistent**: Tends to be an effective listener and can usually gain endorsement through good diplomatic skills. Prefers to motivate by example rather than authority. Tends to like well-defined tasks. May be uncomfortable directing others
  - b. **Efficient, cooperative, realistic:** Tends to generate confidence and trust through persuasion rather than force. Can be direct and assertive to maintain support. Generally supportive and cooperative. Most effective in situations needing a moderate, conservative stance
  - c. **Supportive, cooperative, informal:** Tends to have strong social drives. May depend on feelings more than fact or necessity when making decisions. Prefers to deal with present day issues. May be seen as too trusting and accepting. May not be able to confront tough issues.

- **Driver**
  a. **Controlling, task-oriented, formal:** Strives for accuracy and dependability in performance. A situation that depends on technical expertise satisfies control needs. Leads by example. Analyzes efforts rather than feelings. May downplay relationships.
  b. **Purposeful, businesslike, moderate:** Works best in a well-defined situation in which they determine the course to follow. Likes to be viewed as an authority **figure, but prefers "win-win."** May be seen as single-minded. Tends to rely on power and **position.**
  c. **Competitive, takes charge, task-oriented:** Prefers to deal with immediately relevant issues. Will define goals along with a means of reaching them. Values and rewards loyalty among team members. Achievement oriented.
  d. **Straightforward, aggressive, competitive**: Comfortable in directing others. Tends to influence by personal force. Often perceived as single-minded and demanding. When a situation is consistent with goals, will support a team effort. Recognition and praise are motivators.
- **Expressive**
  a. **Strategic, enthusiastic, practical:** Usually concerned with the importance of relationships. Willing to listen to the other person's point of view. Prone to influence through persuasion rather than force. May rely too heavily on persuasive skills.
  b. **Outspoken, outgoing, stimulating:** Competition and involvement are key attributes. They are persuasive competitors. Will put forth a strong effort to achieve recognition and reward. Won't hesitate to express opinions strongly. May be overly aggressive.
  c. **Enthusiastic, visionary supportive:** Sensitive to the importance of relationships. Can deal with a variety of views tactfully, particularly in a group setting. A team player. Skillful at getting consensus decisions. Responds well to input from others.
  d. **Spontaneous, enthusiastic, futuristic:** Tends to be strongly extroverted. Will emphasize interaction and

involvement when working toward a goal. Often focuses on high-visibility tasks directed at a future goal. Can elicit a strong response, particularly in stressful situations.

INFLUENCING PEOPLE OF OTHER STYLES

- **To Influence the ANALYTIC**
  a. Present facts and data, not feelings or hype.
  b. Emphasize statistical or comparative data.
  c. Be systematic and organized.
  d. Show both sides of the situation/proposal – both the advantages and disadvantages, preferably in chart form.
  e. To minimize the risk, emphasize guarantees or contingencies if "Plan A" doesn't work.
  f. Present your idea/proposal in writing to support your discussion.
  g. Allow him time to think -so he can be cautious, methodical and meticulous.
- **To Influence the AMIABLE**
  a. Present your ideas in a calm, low-key manner.
  b. Show your sincere interest in him as a person.
  c. Match his/her pace – don't rush or pressure for quick decisions.
  d. Demonstrate that you are listening and that you are open to his ideas.
  e. Listen for and discuss how he feels about ideas – support his feelings.
  f. Emphasize any positive effects of your idea/proposal on the people or work groups involved.
- **To Influence the DRIVER**
  a. Avoid small talk
  b. Get to the point quickly
  c. Be brief
  d. Be confident, firm and direct
  e. Be prepared with recommendations
  f. Ask direct questions and wait for a response

g. Respond to questions/challenges in a straight-forward way
h. Confident and succinct manner
i. If you disagree, argue facts, not feelings
j. Keep the relationship businesslike

- **To Influence the EXPRESSIVE**

a. Stress what's unique or new
b. Allow them to talk – make it a conversation
c. Instead of disagreeing, explore options
d. Show some enthusiasm for the topic or request
e. Use testimonials and references from well-known
f. Authorities
g. Don't hurry the discussion – develop stimulating ideas/ approaches together
h. Invited him to talk about his goals or ideas
i. See if/how you can support them and still accomplish

Advantages and Limitations of Communication Models

Model	Communication Example	Advantages	Limitation
Linear	Television and radio E-mail and texts Packaged presentations that do not allow for modifications	**Simple and direct**	Not useful for most face-to-face encounters
Interaction	Instant messaging Class presentations where content is adjusted based on feedback	**Wider applicability**	Still discounts receiver's active role in creating meaning
Transactional	Any encounter in which meaning is concreted	Most realistic depiction of interpersonal communication	Does not apply to texting, tweeting, and posting

FIVE CHARACTERISTICS OF INTERPERSONAL COMMUNICATION

- **Interpersonal communication is a dynamic process:** By dynamic process we mean that interpersonal communication is ongoing or continuous, and in a constant state of flux. All the components continually interact with and affect each other. They are interdependent or interconnected. They depend on and influence one another. What one person says or does influences what the other person says or does. Every interpersonal encounter is a point of arrival from a previous encounter and a point of departure for a future encounter.
- **Interpersonal communication is unrepeatable:** Every interpersonal contact is unique. It has never happened in just that way before, and it will never happen in just that way again. Why? Because every contact changes us in some way and, as a result, can never be exactly repeated or replicated. Try as we might, we can never recapture exactly the same feelings, thoughts, or relationship that existed at a specific point in time. We are no longer exactly the same.
- **Interpersonal communication is irreversible:** In addition to being unrepeatable, interpersonal communication is irreversible. Once we have said or done something to another, we cannot erase its impact. After exhibiting behavior, we cannot simply say, "Forget that!" and substitute a better or more appropriate behavior in its place (though we sometimes would like to try). We cannot rewind or restart communication as we can a TV program recorded on a DVR. We cannot unheard words, unseen sights, or undo acts. They are irretrievable. Presenting a new stimulus does not change the previous stimulus. It merely becomes part of a behavioral sequence.
- **Interpersonal communication is learned:** Over time we learn what works for us in an interpersonal relationship and what does not. We can hinder our communication with another person if we remain unconscious of how we affect him or her, and vice versa. Part of the art of interpersonal communication involves recognizing how our words and actions affect others, how their words and actions affect us, and then, based on our observations, making the necessary adjustments.

5. **Interpersonal communication is characterized by wholeness and nonsummativity:** When we say that the interpersonal relationship is characterized by "wholeness," we are saying that it operates as a complete entity. We consider more than the individuals who are in the relationship; we look at the unique ways in which the persons involved influence each other. When we say that interpersonal communication is characterized by "non summativity," we are saying that the whole is more than the sum of its parts, interpersonal communication is about more than just its participants per se. We cannot understand a couple by looking at only one-half of the partnership. We cannot understand a family by looking at only one of the children. The nature of the relationship must be examined. The "us" must be explored. The relationship takes on a quality that we cannot understand merely by possessing information about its parts. The system as a whole is simply different from the sum of its parts.

TRANSACTION ANALYSIS

- Transaction analysis (T.A.) is a method of understanding communication between people;
- T.A. is a system of analysing and understanding human relationships;
- Transactional analysis was first developed by an American psychiatrist, Eric Berne, drawing on the theories of psychoanalyst Sigmund Freud.

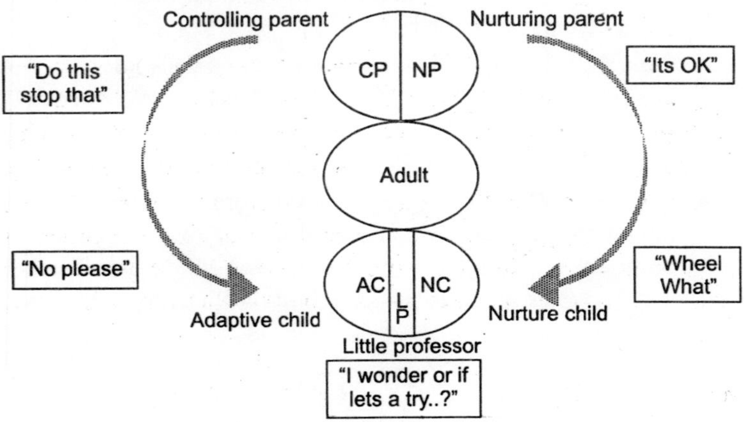

- Eric Berne has developed a very interesting model of interpersonal behaviour that is based on the idea that we have three different states: parent, adult or child. (Sources: Games People Play, Eric Berne, Grove Press 1964 and I'm OK – You're OK, Thomas A Harris, Harper & Row, 1969)

The three different states apply to all of us and are not to confused with age or stage in life. Parent is about a controlling and caring behaviour, while adult is all about cool rationality. The child represents our tendency to emotional reaction and our ability to manipulate emotional responses in others.

The Johari Window

- Graphical model of awareness in interpersonal relationship
- Developed by Joseph Luft and Harry Ingham

	Known to self	**Not known to self**
Known to others	Arena or free activity	Blind spot
Not known to others	Façade or Hidden Aras	Unworn

- **Quadrant I (Arena):** This refers to those items about ourselves which we know about, are willing to expose to others and get feedback on. Examples could include skills such as those relating to foreign languages.
- **Quadrant II (Blind spot):** This is sometimes called the 'bad breath syndrome'. Others know about this feature but we are unaware of it. Sadly we get no feedback and remain ignorant of an interpersonal weakness that actually holds back out relationships with others.
- **Quadrant III (Façade or hidden areas):** This refers to those features which we know about ourselves but we choose not to reveal to others. Examples here would be a weakness about which we are embarrassed, such as an inability to dance.
- **Quadrant IV (Unknown):** This covers items that we may not be aware of and which others are not aware of either. The point about willing to expose ourselves and to gain feedback is to move as much as possible into the 'healthy' arena area.

CHAPTER 10

CONFLICT MANAGEMENT

Conflict is an outcome of behaviors. It is an integral part of human life. Whenever there is interaction there is conflict. Conflict means expression of hostility, negative attitudes, antagonism, aggression, rivalry, and misunderstanding. Conflict means to be in opposition to one another. It refers to disagreement between people or members of organizations. Conflict is difficult to define, because it occurs in many different settings. The essence of conflict seems to be disagreement, contradiction, or incompatibility. Thus, conflict refers to any situation in which there are incompatible Goals, Cognitions, or Emotions within or between individuals or groups that lead to opposition or antagonistic interaction.

THE DEFINITION RECOGNIZES THREE BASIC TYPES OF CONFLICT

- **Goal conflict** is situation in which desired end states or preferred outcomes appear to be incompatible.
- **Cognitive conflict** is a situation in which ideas or thoughts are inconsistent.
- **Affective conflict** is a situation in which feelings or emotions are incompatible; that is, people literally become angry with one another. Conflict is very common in organizational settings. This is not necessarily a negative feature; the resolution of conflict often leads to constructive problem solving.

LEVELS OF CONFLICT

Intrapersonal Conflict

Intrapersonal conflict, which occurs within an individual, often involves some form of goal conflict or cognitive conflict. Goal conflict exists for individuals when their behaviour will result in outcomes that are mutually exclusive or have compatible elements (both positive and negative outcomes).

- Approach-approach conflict is a situation in which a person has a choice between two or more alternatives with positive outcomes; for example, a person can choose between two jobs that appear to be equally attractive.
- Avoidance - avoidance conflict is a situation in which a person must choose between two or more alternatives, and they all have negative outcomes. For example, employees may be threatened with punishment in the form of demotion unless they do something they dislike spend much time travelling on their job, for example.
- Approach-avoidance conflict is a situation in which a person must decide whether to do something that had both positive and negative outcomes, for example, being offered a good job in a bad location.

Interpersonal Conflict

Interpersonal conflict involves two or more individuals rather than one individual. Two managers competing for the same promotion, two executives maneuvering for a larger share of corporate capital examples of conflict between individuals are legion and quite familiar.

Reasons

- **Personality differences**: Some people have difficulty in getting along with each other. This is purely a psychological problem and it has nothing to do with their job requirements or formal interactions.
- **Perceptions:** Varied backgrounds, experiences, education and training result in individuals developing different perceptions of similar realities; te result being an increase in the likelihood of interpersonal conflict.
- **Clashes of values and interests:** Conflict that so commonly develops between engineering and manufacturing personnel shows how differences in values might underlie conflict. Members of the engineering department might place a

premium on quality, sophisticated design and durability while members of the manufacturing department might value simplicity and low manufacturing costs.

- **Power and status differences:** As pointed out by Abraham Zalenznik "Organizations are political structures". They operate by distributing authority and setting a stage for the exercise of power. Similarly status inconsistencies lead to conflict.
- **Scarce resource:** Interpersonal conflict is almost automatic anytime there is scarcity. Conflicts over scarce resources are exceedingly common in organizations. Where the scarcity is absolute (the resource level cannot be enhanced) it is very difficult to manage.

Intragroup Conflict

A group experiencing intragroup conflict, may eventually resolve it, allowing the group to reach a consensus. Or the group may not resolve the conflict, and the group discussion may end in disagreement among the members. A study of a large number of groups engaged in business and governmental decision making, tried to identify some the conditions that lead to (1) the successful resolution of conflict (consensus or (2) the failure to resolve conflict (disagreement). This study showed that conflict within groups is not a simple, single phenomenon. Instead, intragroup conflict seems to fall into two distinct categories : (1) substantive conflict and (2) affective conflict. Substantive conflict refers to conflict based on the nature of the task or to "content" issues. It is associated with intellectual disagreements among the group members. In contrast, affective conflict derives primarily from the group's interpersonal relations. It is associated with emotional responses aroused during interpersonal clashes.

Inter-Group Conflict

An organization is a collection of individuals and groups. As the situation and requirements demand, the individuals form various groups. The success of the organization as a whole depends upon the harmonial relations among all interdependent groups, even though some intergroup conflicts in organizations is inevitable. The idea is to study intergroup behaviours within an organization so that any conflict can be recognized and dealt with by the management.

Intra-Organizational Conflict

Four types of intra-organizational conflict exist:

- **Vertical conflict:** Vertical conflict refers to any conflict between levels in an organization; superior-subordinate conflict is one example. Vertical conflicts usually arise because superiors attempt to control subordinates and subordinates.
- **Horizontal conflict:** Horizontal Conflict refers to conflict between employees or departments as the same hierarchical level in an organization.
- **Line-Staff conflict:** Most organizations have staff departments to assist the line departments. The line-staff relationship frequently involves conflict. Staff managers and line managers typically have different personal characteristics. Staff employees tend to have a higher level of education, come from different backgrounds, and are younger than line employees.
- **Role conflict:** A role is the cluster of activities that others expect individuals to perform in their position. A role frequently involves conflict.

Managing Conflict

Preventing conflict. Some of the preventive measures that the management can take, **according to Schein** are discussed below:

- **Goal structure:** Goals should be clearly defined and the role and contribution of each unit towards the organizational goal must be clearly identified. All units and the individuals in these units must be aware of the importance of their role and such importance must be fully recognized.
- **Reward system:** The compensation system should be such that it does not create individual competition or conflict within the unit. It should be appropriate and proportionate to the group effort and reflect the degree of interdependence among units where necessary.

- **Trust and communication:** The greater the trust among the members of unit, the more honest and open the communication among them would be. Individuals and units should be encouraged to communicate openly with each other so that they can all understand each other, understand each other's problems and help each other when necessary.
- **Co-ordination:** Co-ordination is the next step to communication. Properly co-ordinated activity reduce conflict. Wherever there are problems in co-ordination, a special liaison office should be established to assist such co-ordination.

RESOLVING BEHAVIOURAL CONFLICT

- **Ignoring the conflict:** In certain situations, it may be advisable to take a passive role and avoid it all together. From the manager's point of view, it may be especially necessary when getting involved in a situation would provoke further controversy or when conflict is so trivial in nature that it would not be worth the manager's time to get involved and try to solve it. It could also be that the conflict is so fundamental to the position of the parties involved that it may be best either to leave it to them to solve it or to let events take their own course. The parties involved in the conflict may themselves prefer to avoid conflict, specially, if they are emotionally upset by the tension and frustration created by it. People may intrinsically believe that conflict is fundamentally evil and its final consequences are never good. Thus people may try to get away from conflict causing situations.
- **Smoothing:** Smoothing simply means covering up the conflict by appealing for the need for unity rather than addressing the issue of conflict itself. An individual with internal conflict may try to count his blessings and forget about the conflict. If two parties have a conflict within the organization, the supervisor may try to calm things down by being understanding and supportive to both parties and appealing them for cooperation. The supervisor does not ignore or withdraw from the conflict nor does he try to address and solve the conflict but expresses hope that "everything will work out for the best of all." Since

the problem is never addressed, the emotions may build up further and suddenly explode. Thus smoothing provides only a temporary solution and conflict may resurface again in the course of time. Smoothing is more sensitive approach than avoiding in that as long as the parties agree that not showing conflict has more benefits than showing conflicts, the conflict can be avoided.

- **Compromising:** A compromise in the conflict is reached by balancing the demands of the conflicting parties and bargaining in a give and take position to reach a solution. Each party gives up something and also gains something. The technique of conflict resolution is very common in negotiations between the labour unions and management. It has become customary for the union to ask for more than what they are willing to accept and for management to offer less than what they are willing to give in the initial stages. Then through the process of negotiating and bargaining, mostly in the presence of arbitrators, they reach a solution by compromising. This type of compromise is known as integrative bargaining in which both sides win in a way.
- **Forcing:** As Webber puts it, "the simplest conceivable resolution is the elimination of the other party – to force opponents to flee and give up the fight – or slay them." This is technique of domination where the dominator has the power and authority to enforce his own views over the opposing conflicting party. This technique is potentially effective in situations such as a president of a company firing a manager because he is considered as a trouble-maker and conflict creator. This technique always ends up in one party being a loser and the other party being a clear winner. Many professors in colleges and universities have lost promotions and tenured re-appointments because they could not get along well with their respective chairpersons of the departments and had conflicts with them. This approach causes resentment and hostility and can backfire. Accordingly, management must look for better alternatives, if these become available.

- **Problem solving:** This technique involves confronting the conflict in order to seek the best solution to the problem. This approach objectively assumes that in all organizations, no matter how well they are managed, there will be differences of opinions which must be resolved through discussions and respect for differing viewpoints. In general, this technique is very useful in resolving conflicts arising out of semantic misunderstandings. It is not so effective in resolving non-communicative types of conflicts such as those that are based on differing value systems, where it may even intensify differences and disagreements. In the long run, however, it is better to solve conflicts and take such preventive measures that would reduce the likelihood of such conflicts surfacing again.
- **Collaborating:** Incorporates multiple viewpoints and insights from differing perspectives; leads to consensus and commitment.
- **Confronting:** Confronting style refers to strong cooperative and assertive behaviour. It is the win-win approach to interpersonal conflict handling. The person using confronting desire to maximize joint results. An individual who uses this style tends to see conflict as natural, helpful, and leading to a more creative solution if handled properly. Exhibit trust in others and conflict is resolve to the satisfaction of all. Confronting style is most practical to find an integrative solution when both sets of concerns are too important to be compromised, when objective is to learn, to gain commitment by incorporating concerns into a consensus and to work through feelings that have inferred with a relationship.

Conflict resolution model based on New storm and Davis

CHAPTER 11

TIME MANAGEMENT

DEFINITION OF TIME MANAGEMENT

Time, can be considered as a series of "NOWS". The only time you have controlled over or when you really have to do anything is "NOW".

Management, process consist of the functions of planning, organizing, motivating and controlling.

TIME MANAGEMENT

- It is the skill of taking action for the right reason at the right time in the right way.
- It is the process of planning and exercising conscious control over the amount of time spent on specific activities - especially to increase effectiveness, efficiency or productivity.
- Time management is a process, which consists of the functions of planning, organizing, motivating and controlling the resource time.
- Time management is a skill, an attitude and a technique which one can learn and develop.

TYPES OF TIME

- **Biological time:** This is the time we spend on sleep, physical recreation and food etc.
- **Social time:** This is the time we spend on meeting friends, making social calls, talking on the telephone, family occasions, get-togethers, spend on self-development like reading etc.
- **Business time:** This is the time spent on meetings, interviews, data entry etc.

Myths About Time Management

- The harder one works, the more he gets done.
- People in higher positions make better decisions.
- Delay improves the quality of decisions.
- Most people are overworked because of the nature of their job.
- Time can be saved.
- If you do things yourself, tasks get done more efficiently.
- Delegation would save you from worry and responsibility.
- Efficiency means effectiveness.
- It is not necessary to write out your objectives.
- For busy people , spending time in planning the day is a waste of time.
- All activities lead to goal achievement.
- Busy people perform more than others.
- Most people know how to spend time.
- Planning of only official time is needed.
- Indians are bad time managers.

Time Management Issues

- **Shifting priorities and crisis management:** Management guru Peter Drucker says that "crisis management is actually the form of management preferred by most managers" The irony is that actions taken prior to the crisis could have prevented the fire in the first place.
- **The telephone:** Have you ever had one of those days when you thought your true calling was in telemarketing? The telephone-our greatest communication tool can be our biggest enemy to effectiveness if you don't know how to control its hold over you.
- **Lack of priorities/objectives:** This probably the biggest/ most important time waster. It affects all we do both professionally and personally. Those who accomplish the most in a day know exactly what they want to accomplish. Unfortunately too many of us think that goals and objectives are yearly things and not

daily considerations. This result in too much time spent on the minor things and not on the things which are important to our work/lives.

- **Attempting too much:** Many people today feel that they have to accomplish everything yesterday and don't give themselves enough time to do things properly. This leads only to half-finished projects and no feeling of achievement.
- **Drop in visitors:** The five deadliest words that rob your time are "Have you got a minute". Everyone's the culprit-colleagues., the boss, your peers. Knowing how to deal with interruptions is one of the best skills you can learn.
- **Ineffective delegation:** Good delegation is considered a key skill in both managers and leaders. The best managers have an ability to delegate work to staff and ensure it is done correctly. This is probably the best way of building a team's moral and reducing your workload at the same time. The general rule is -this; if one of your staff can do it 80% as well as you can, and then delegate it.
- **The cluttered desk:** When you have finished reading this article look at your desk. If you can see less than 80% of it then you are probably suffering from 'desk stress'. The most effective people work from clear desks.
- **Procrastination:** The biggest thief of time; not decision making but decision avoidance. By reducing the amount of procrastinating you do, you can substantially increase the amount of active time available to you?
- **The inability to say "no!":** The general rule is; if people can dump their work or problems on to your shoulders they will do it. Some of the most stressed people around lack the skill to 'just say no' for fear of upsetting people.
- **Meetings:** Studies have shown that the average manager spends about 17 hours a week in meetings and about 6 hours in the planning time and untold hours in the follow up. I recently spoke to an executive who has had in the last 3 months 250 meetings It is widely acknowledged that about as much of a third of the time spent in meetings is wasted due to poor meeting management and lack of planning. If you remember

your goal is to increase your self management, these are the best ways to achieve this.

Importance of Time Management

- **Time is limited:** No matter how you slice it, there are only 24 hours in a day. That applies to you, and to your coworker who only seems able to do half the amount of work you do. But it also applies to the former coworker who consistently accomplishes more than you, and was promoted as a result. If you want to rise through the ranks, you have to acknowledge the importance of finding a way to manage this limited resource.
- **Time is a special resource:** That you cannot store or save for later use. Everyone has the exact same amount of time each day. Time not well used cannot be retrieved.
- **You can accomplish more with less effort:** When you learn to take control of your time, you improve your ability to focus. And with increased focus comes enhanced efficiency, because you don't lose momentum. You'll start to breeze through tasks more quickly (the workday will also seem to fly by).
- **Improved decision-making ability:** Whether you rely on a time-chunking technique or discover the power of list-making, you'll soon find that a nice side benefit of good time management skills is the ability to make better decisions. When you feel pressed for time and have to make a decision, you're more likely to jump to conclusions without fully considering every option. That leads to poor decision making.
- **Become more successful in your career:** Time management is the key to success. It allows you to take control of your life rather than following the flow of others. As you accomplish more each day, make more sound decisions, and feel more in control, people notice. Leaders in your business will come to you when they need to get things done. And that increased exposure helps put you in line for advancement opportunities.
- **Learning opportunities are everywhere:** Obviously, the more you learn, the more valuable you are to your employer.

And great learning opportunities are around you, if you've got time to stop and take advantage them.

- **Reduce stress:** When you don't have control of your time, it's easy to end up feeling rushed and overwhelmed. And when that happens, it can be hard to figure out how long it's going to take to complete a task. (Think of a time when you were about to miss a deadline and were frantically trying to finish the project. If someone dumped a surprise on your desk at that moment and asked you how long it would take to finish the surprise task, how could you even begin to answer their question?)
- **Free time is necessary:** Everyone needs time to relax and unwind. Unfortunately, though, many of us don't get enough of it. Between jobs, family [illegible] ties, errands, and upkeep on the house and the [illegible] us are hard-pressed to find even 10 minutes to sit and do nothing.
- **Self-discipline is valuable:** When you practice good time management, you leave no room for procrastination. The better you get at it, the more self-discipline you learn. This is a valuable skill that will begin to impact other areas of your life where a lack of discipline has kept you from achieving a goal.
- **Time management** will help you set up your priorities.
- **Get more done in less time:** The other reason why time management is important is that you will get more done in less time. When you plan your time, you will see the amount of work that needs to be tackled. That puts into context how much time you need. Inevitably, you will know not to laze your time away.
- **More quality work:** When you plan your time you are making wise investment of a very limited resource – time. Knowing what to work on, when and how much time you have to finish the work makes you more focused. That focus on the work ensures that you get more quality work out.
- **Keeps things in context:** Sometimes you may hear people complaining that there is no balance in their lives.

- **Keeps you on track:** Whether it is a daily work list or a long-term career plan – time management keeps you on track. With your goals stated and time allocated, the likelihood of you staying within the boundaries you have given yourself is higher. Of course, do not be your own jury and judge. That means do not give yourself excuses when you start to procrastinate.
- **Making sure you deliver what's promised:** How would you deliver what you promised on time if you do not manage your time? Have you always disappointed your colleagues on the delivery of your part of the project? Perhaps you aren't doing well with time management. Making sure you deliver what's promised should be reason enough why time management important is.

Time Management Tips

The following tips can help you stay on track:

- **Conquer the clutter:** Schedule 10 to 15 minutes each week to clear your work area of junk mail, old papers, and other accumulated clutter. Change habits that lead to messes. Keep cleaning supplies handy so you can take advantage of the odd free moment to police your work space.
- **Defuse distractions:** Little distractions can add up to a major drain on productivity. If you're spending too much time on the phone, keep an egg timer at your desk and hold calls to a reasonable limit. Learn how to terminate calls politely. If co-workers often drop in to chat, close your door. If you're constantly walking around obstacles, consider a change of floor plan. Take steps to reduce distracting noise.
- **Know the self:** Rivers can't be forced to flow uphill; nor should you try to work against your inner nature. Schedule the toughest work for your circadian period of peak productivity. Minimize the impact of suboptimal climate control with a fan or small heater. Perhaps better or different lighting would boost your efficiency.
- **Eliminate redundancy:** Analyze every process you use to determine if any steps can be eliminated. Common problem include multiple signatures for approval, extra steps designed

to circumvent systems or correct problems that could be addressed more directly, and generating multiple copies that are no longer required.

- **Group and separate:** Tedious or redundant tasks can be grouped for increased efficiency: file all at once, bill all at once, order all at once. Large, multifaceted tasks, on the other hand, may be best tackled in small pieces. For example, sort that large stack of paperwork on Monday, process some on Tuesday, some on Wednesday, and so forth until it's done. Using this approach, even the most daunting tasks become manageable.
- **Share the burden:** As any quilter knows, many hands make light work, especially when tackling tedious or large tasks. Performance of dreaded chores like the annual inventory can take on a party atmosphere when many are involved and frequent breaks are scheduled.
- **Seize the moment:** We all have a tendency to put off minor, less important tasks, and we also spend significant chunks of time holding on the phone or waiting in line. Can you see an opportunity here? Make a list of tasks that take five, 10, or 20 minutes, and keep the materials you need to do these tasks handy. That way, when you're put on hold or stuck in line, you can pull out that small job and finish it up.
- **Emulate others:** Don't try to reinvent the wheel. If someone else always seems to be ahead of the game, watch and learn. If someone else has a speedier way of doing something, copy it. If you're having trouble getting specific jobs done, ask others how they organize and execute the task; perhaps you've overlooked some short cuts.
- **Make work fun:** Introducing a bit of fun into your work will make the day easier for you and your customers. Challenge yourself to process one hundred pieces of paper every day for five days. Have a friendly contest with a co-worker to see who can process the most billing statements in an hour with no mistakes. If possible, flip your morning and afternoon schedules for a change of pace.

Time Management Strategies

- **Pareto analysis:** This is the idea 80% of tasks can be completed in 20% of the disposable time. The remaining 20% of tasks will take up 80% of the time. This principle is used to sort tasks into two parts. According to this form of Pareto analysis it is recommended that tasks that fall into the first category be assigned a higher priority. The 80-20-rule can also be applied to increase productivity: it is assumed that 80% of the productivity can be achieved by doing 20% of the tasks. Similarly, 80% of results can be attributed to 20% of activity. If productivity is the aim of time management, then these tasks should be prioritized higher it depends on the method adopted to complete the task. There is always a simpler and easier way to complete the task. If one uses a complex way, it will be time consuming. So, one should always try to find out alternative ways to complete each task.
- **The Eisenhower method:** The "Eisenhower Method" stems from a quote attributed to Dwight D. Eisenhower: "I have two kinds of problems, the urgent and the important. The urgent are not important, and the important are never urgent." Using the Eisenhower Decision Principle, tasks are evaluated using the criteria important/unimportant and urgent/not urgent and then placed in according quadrants in an Eisenhower Matrix (also known as an "Eisenhower Box" or "Eisenhower Decision Matrix").

	Urgent	**Not urgent**
Important	Crises deadlines problems	Exercises Vocation Planning
Not important	Interruption Distraction Other calls	Trivia Busy work Time wasters

a. **Important/urgent quadrant** are done immediately and personally e.g. crises, deadlines, problems.

b. **Important/not urgent** quadrant get an end date and are done personally e.g. relationships, planning, recreation.

c. **Unimportant/urgent quadrant** are delegated e.g. interruptions, meetings, activities.

d. **Unimportant/not urgent quadrant** are dropped e.g. time wasters, pleasant activities, trivia.

- **Domino reaction method:** This is the idea that there are actions that you invest in once and which produce over time in different channels. Writing a book is such an action, because it requires a one-time effort, and once you finish it, it continues serving you.
- **POSEC method:** POSEC is an acronym for **Prioritize by Organizing, Streamlining, Economizing and Contributing**. The method dictates a template which emphasizes an average individual's immediate sense of emotional and monetary security. It suggests that by attending to one's personal responsibilities first, an individual is better positioned to shoulder collective responsibilities.

a. **Prioritize:** Your time and define your life by goals.

b. **Organize:** Things you have to accomplish regularly to be successful (family and finances).

c. **Streamline:** Things you may not like to do, but must do (work and chores).

d. **Economize:** Things you should do or may even like to do, but they're not pressingly urgent (pastimes and socializing).

e. **Contribute:** By paying attention to the few remaining things that make a difference (social obligations).

CHAPTER 12
PUBLIC SPEAKING

****We have nothing to fear, but fear itself*

(Franklin D. Roosevelt) ****

INTRODUCTION

Speaking is an important method for communicating knowledge and expressing ideas. Effective and excellent communication skill is an advantage for work in school, business as well as your personal life. Delivering an effective presentation to 20 or to 200 people is difficult. Since, listeners have better access to information owing to advent of the Internet audiences expect more content from speakers today. In addition, because of the entertainment slant of most media today, audiences want a presentation delivered with animation, humor and excitement. Public speaking can serve the purpose of transmitting information, telling a story, motivating people to act or some combination of those. Public speaking can also take the form of a discourse community, in which the audience and speaker use discourse to achieve a common goal. Public speaking for business and commercial events is often done by professionals. These speakers can be contracted independently, through representation by a speaker's bureau, or by other means. Public speaking plays a large role in the professional world; in fact, it is believed that 70 percent of all jobs involve some form of public speaking.

- Public speaking is the process and act of speaking or giving a lecture to a group of people in a structured, deliberate manner intended to inform, influence, or entertain a listening audience. Public speaking is commonly understood as face-to-face speaking between individuals and an audience for the purpose of communication. It is closely allied to "presenting", although the latter is more often associated with commercial activity. Most of the time, public speaking is to persuade the audience (source Wikipedia).

- Public speaking (also called oratory or oration) is the process or act of performing a speech to a live audience.
- Public speaking is commonly understood as formal, face-to-face speaking of a single person to a group of listeners.

Basic Elements of Public Speaking

- Communicator,
- Message,
- Medium,
- Audience
- Effect.

Tools of Public Speaking

- Public address system
- Telecommunication
- Videoconferencing,
- Social media

Types of Public Speaking

- **Speaking to inform:** When a person gives a speech before an audience to impart information on a particular topic or issue it said to be an informative speech. Business presentations, seminars in colleges, class presentations in schools are some examples of informative speeches. A person preparing for an informative speech has to research the subject or topic very well. It should be short and precise because long informative speeches (lectures) easily bore the audience. The success of an informative speech will depend on how much the audience could understand from the speech.
- **Speaking to persuade:** Persuasive speeches are those where you try to persuade or convince your audience about an idea or product. These speeches aim to influence and change the opinions of the audience. This can be a difficult task as you

could be facing a group of people who may have totally different views from your own. The most important point that has to be kept in mind here is that if you want to influence others' views and ideas, you have to show your enthusiasm while speaking.

- **Speaking to actuate:** Speaking to actuate is a higher level of persuasive speaking. Here, the speaker goes a step beyond persuasion and convincing. The aim is to motivate people enough to take a specific step—to act. This is a powerful level of speaking. Very few people have achieved this level of mastery of the art of persuasion where they could so deeply convince people that they move into action. This type of speech is useful in conflict situations like war. It is also in action when a speaker not only tries to garner support for some cause or reason but also wants the people he's speaking to join him in actualizing it through concrete action. It is characterized by a show of charisma, very strong words, and great emotional involvement and display of conviction.
- **Speaking to entertain:** Ceremonial speeches are another form of public speaking usually given at weddings, funerals, graduation parties, retirement parties etc. One very important factor to make these speeches effective is to add a personal touch. One gives these speeches for people one knows and it would be great if one can bring in stories and incidents about the respective person. The speech can be humorous, touching or emotional, as per the occasion and the mood. However, one should take utmost care not to hurt feelings by making snide remarks about them.

Voice Modulation

Voice modulation is an important aspect of effective public speaking. Without voice modulation, the speech would sound flat and will not be able to catch audience attention. The time span for the attention of the audience is very short so the speaker has to draw the interest of audience so that they listen to him carefully. Voice modulation is necessary to engage your listener.

- **Tone:** The tone of voice conveys about the situation whether a serious or joyous situation. Tone helps in visualizing the words of the speaker. It's very much necessary to speak in the right tone and stress words which carry importance. You need to take care as to when you should change the pitch to stress a word.
- **Pause:** Pause helps in understanding the meaning of the spoken words. Pause helps in enhancing the meaning of words and it also provides a breathing opportunity.

  How to use the pause effectively:

  a. After you are introduced
  b. When you want to emphasize certain ideas
  c. Pause after questions
  d. Use the pause as a transitional device – When moving from one point to another
  e. Pause to get the best effect from humor – pausing before and after the humor prepares the audience to understand and react to the message.
  f. Pause after the end of the speech. It will give time to the audience to react.
  g. Pause to add your overall effectiveness as a speaker.
- **Silence:** You keep silence so that the audience absorbs what you have said.

Social Graces

Social graces are skills used to interact politely in social situations. They include manners, etiquette (the specifically accepted rules within a culture for the application of universal manners), deportment, fashion and refinement (also known as sophistication). These skills were once taught to young women at a finishing school or charm school. The focus of social graces has changed over the last century, recently with an emphasis on business etiquette and international protocol.

Email Etiquette

- **Answer swiftly:** Your customers' send you email because they want quick responses. The golden rule for email is to reply within 24 hours, and preferably within the same working day. If your response email is complicated, just send an email confirming receipt and letting them know that you will get back to them. This will ease the customer's mind.
- **Use a meaningful subject line:** Try to use a subject that is meaningful to the recipient as well as yourself. For instance, when you send an email about a product, it is better to mention the actual name of the product, e.g. 'Product An information' than to just say 'Product information'. It also makes it easier to search for old emails when the subject line is relevant and specific to the content of the email.
- **Don't abuse the "Reply to All":** Only use Reply to All if you really need your message to be seen by each person who received the original message. Sending off irrelevant or unnecessary replies to everyone on the list is just annoying and confusing. However, if communication is vital between all parties in an email thread, use the Reply to All to keep everyone in the loop. If you only use Reply in such a case, the recipient may have to forward your email to everyone else, which is frustrating and disjointed.
- **Use the BCC Field:** When sending to many people, some people put all the email addresses in the to: field. There are two drawbacks to doing that: (1) the recipient knows that you have sent the same message to a large number of recipients, and (2) you are publicizing someone else's email address without their permission. Instead, consider using the Bcc: field. Put your mailing list group name in to: field in their email (leaving the to: field blank may look like spam). If you have Microsoft Outlook and Word you can do a mail merge so each recipient receives their own email, or create a mailing group in your email software if it has that utility.
- **Don't leave out the message thread:** Include the original mail in your reply, in other words click 'Reply', instead of 'New Mail'. We all receive many emails and we can't

remember each individual email. Leaving the thread may take a fraction longer in download time, but it saves the recipient time looking for the related emails in their inbox. Remember, emails are not like regular printed correspondence - the name of the game is to keep it quick and efficient – so include the thread!

- **Read your email before you send it:** Treat email like any other official company document. Read it before you send it.
- **Spelling and grammar errors** are just as unfortunate in email as anywhere else in your corporate correspondence. Look out for potential misunderstandings, the tone, and inappropriate comments; we use email because it is quick and easy but precisely that quickness may cause more trouble than you bargained for.
- **Confidential information:** Email is just too risky a place to include confidential information. Ask yourself if you would want the content of your email displayed on a bulletin board. Never make libelous, sexist or racially discriminating comments in emails, even as a joke. Consider implementing a Disclaimer on the bottom of all corporate emails with statements on Breach of Confidentiality, Virus Liability, etc. (Yes, you can be sued for sending an email that contains a virus!)
- **Abbreviations & emoticons:** Be careful using email abbreviations such as BTW (by the way) and LOL (laugh out loud) in business emails. Even today, some people still don't know what they mean, so it's better to drop them and emoticons, such as the smiley :-) don't belong in business email unless a relaxed form of communication has long been established with the customer.
- **Don't attach unnecessary files:** Wherever possible try to compress attachments and only send attachments when they are productive. Make sure you have good virus software in place.
- **Don't forward junk:** Don't forward chain letters, virus hoaxes, chain email solicitations for charitable causes even if they sound bonafide, funny pictures and jokes. Would you

put these things on your corporate letterhead? I don't think so. Don't ever send or forward emails containing libelous, defamatory, offensive, racist or obscene remarks. Just one offensive remark can result in a multi-million dollar court case for you and your company.

- **Be concise:** Do not make an email longer than it needs to be. Email is harder to read than printed communications. A long email can be very discouraging and can be abandoned before the recipient gets to your final point all the way down at the bottom. If it has to be long, consider including a synopsis at the top of the email.
- **Answer all questions & more:** Make sure you answer all the questions and pre-empt new questions in your reply. If you don't answer all the questions in the original email, you're wasting your own, your company's and your customer's time. Worse still, you are leaving the customer frustrated. By answering all questions and pre-empting further inquiries, you are making a great impression and reflecting thoughtful customer service. For example, a question regarding the types of credit cards you accept can be replied to with the list of cards, information on other payment methods and even a link to your website order information page.
- **Make it personal:** Did you know that the most effective word in marketing is "you"? Not only should the e-mail be personally addressed, it should also include personal, i.e. customized, content. For this reason auto replies are usually not very effective. When you get some questions over and over, such as directions to your office or how to subscribe to your newsletter, save these texts as response templates and paste them into your message when you need them. You can save your templates in a Word document, or use pre-formatted emails.
- **Use the proper structure & layout:** Reading from a screen is more difficult than reading from paper so the structure and layout is very important for email messages. Make your paragraphs short and use blank lines between each paragraph. When making points, number them or separate each point with blank lines to keep the overview.

- **Don't overuse the high priority function:** We all know the story of the boy who cried wolf. If you overuse the high priority option, it will lose its function when you really need it. Besides, even if a mail has high priority, your message will come across as slightly aggressive if you flag it as 'high priority'. Likewise, be careful using the words Urgent or Important in the subject line.
- **Do not write in CAPITALS:** IF YOU WRITE IN CAPITALS IT SEEMS AS IF YOU ARE SHOUTING!! This can be highly annoying, difficult to read and might trigger an unwanted response in the form of a flame mail (you get yelled back at!). Therefore, try not to send email text in capitals.
- **Be careful with formatting:** Remember that when you use formatting in your emails, the sender might not be able to view formatting, or might see different fonts than you had intended. 10% of email recipients cannot read html or rich text email; they can only receive in plain text. So for them, fonts, colors and other fancy formatting is lost. When using colors, make sure it is easy to read on the background color you have selected. Remember, monitors vary in color presentation so what may look good on your monitor may be unreadable when displayed on another monitor.
- **Do not request delivery & read receipts:** This will almost always annoy your recipient before he or she has even read your message. Besides, it usually does not work anyway since the recipient could have blocked that function, or his/her software might not support it, so what is the use of using it? If you want to know whether an email was received it is better to ask the recipient to let you know that it was received.
- **Do not recall a message:** Chances are that your message has already been read. A recall request just looks silly then. It is better to send an email saying you have made a mistake. This will look much more honest than trying to recall a message.
- **Do not copy a message or attachment without permission:** Do not copy a message or attachment belonging to another user without permission of the originator; you might infringe on copyright laws.

- **Avoid long sentences:** As mentioned earlier, email is harder to read than printed material. People don't give email the same brain power as they do when reading for example a letter. Try to keep your sentences to no more than 15-20 words.
- **Keep your language gender neutral:** It is correct today to avoid sexist language such as: "The customer should bring his car to our service department for an oil change". You can use "his/her" or keep it neutral by rephrasing the sentence: "The customer should bring the car to our service department for an oil change".
- **Don't reply to spam:** Spam may make you furious and you may want to reply with "flame mail". Many spam emails are sent to confirm that your email address is still valid, and by replying you are only making yourself known to dubious marketers which may result in even more spam. Just delete the spam, or use anti-spam software.

Telephone Etiquette

Proper telephone etiquette is very important in that you are representing your department and the University. Remembering to use proper telephone etiquette, whether answering the phone or making phone calls, leaves callers with a favorable impression of you, your department, and Lehigh in general.

- **Presenting a professional image:** Both in person and on the telephone, is very important in the Office skills profession. Taking care of your customers over the telephone and making them feel well informed and appreciated is essential. The ability to listen is a problem in general but it is very important to listen to what the caller has to say. It is always a good habit to repeat the information back to the client when you are taking a message.
- **Speaking style:** A picture paints a thousand words but the caller on the other end of the phone can only hear you. Caller cannot see your face or body language. Taking the time to speak clearly, slowly and in a cheerful, professional voice is very important.

- **Usage of tone:** Use your normal tone of voice when answering a call. In case if you have a tendency to speak loud or shout, avoid doing so on the telephone.
- **Usage of language:** Speak in a normal way but in an effective manner. Never use swear words. Words delivered should be clear & understood to the caller.

Do's

- Be patient and helpful.
- Respond clearly with "yes" or "no" when speaking.
- Always remember it is important
- If you are answering a call on behalf of your team member please mention to the caller & try if you can assist them on the query.
- Else note down the caller's number so that a concerned user can make a call back on the query for resolution.
- Address the caller properly by his or her title.
- Eg: (Good morning Mr. Brown, Good afternoon Ms. Sanders).
- Listen to the caller and what they have to say.
- Always ask if you can put the caller on hold.
- Verify that you have heard and transcribed the message accurately

Don'ts

- Do not eat or drink while you are on telephone duty.
- Do not use slang words or poor language.
- Never address an unfamiliar caller by his or her first name.
- Never snap back or act rude to the caller.
- Never leave the person on hold for more than a few seconds or they may become upset and hang up.

Answering the call

- Good morning/afternoon/evening. Thank you for calling [department/floor]. This is [your name]. How may I assist you?

- Answer call with smile
- Keep voice tone warm and pleasant
- Speak clearly and concisely
- Once you get the caller's name -use it
- Give full attention
- Do not become distracted with other tasks
- **Ending Call:** Thank you for calling [caller's name]. Is there anything else I can do for you today?"
- **On Hold:** "[caller's name], may I place you on a brief hold? " Check on caller every 2-3 minutes. Max hold time 6 minutes. If you have to put caller on hold twice, obtain contact name, number, reason.
- **Transfers**
  a. Only transfer if you know who the appropriate person is who can help the caller.
  b. Explain why you are transferring the call.

     If you are asked a question to which you do not know the answer.
  c. Take name, number and question.
  d. Find out answer, return call.

CHAPTER 13

COLLECTIVE BARGAINING AND NEGOTIATION SKILLS

DEFINITION

- Collective bargaining is a type of negotiation used by employees to work with their employers.
- During a collective bargaining period, workers' representatives approach the employer and attempt to negotiate a contract which both sides can agree with.
- Typical issues covered in a labor contract are hours, wages, benefits, working conditions, and the rules of the workplace.
- Once both sides have reached a contract that they find agreeable, it is signed and kept in place for a set period of time, most commonly three years.
- The final contract is called a collective bargaining agreement, to reflect the fact that it is the result of a collective bargaining effort.

FEATURES

- It is a collective process.
- The representatives of both workers and management participate in bargaining.
- It is a continuous process. It establishes regular and stable relationship between the parties involved. It involves not only the negotiation of the contract, but also the administration of the contract.
- It is a flexible and dynamic process. The parties have to adopt a flexible attitude through the process of bargaining.
- It is a method of partnership of workers in management.

IMPORTANCE TO SOCIETY

- Collective bargaining leads to industrial peace in the country.
- It results in establishment of a harmonious industrial climate which supports, which helps the pace of a nation's efforts towards economic and social development since the obstacles to such a development can be reduced considerably.
- The discrimination and exploitation of workers is constantly being checked.
- It provides a method or the regulation of the conditions of employment of those who are directly concerned about them.

IMPORTANCE TO EMPLOYERS

- It becomes easier for the management to resolve issues at the bargaining level rather than taking up complaints of individual workers.
- Collective bargaining tends to promote a sense of job security among employees and thereby tends to reduce the cost of labor turnover to management.
- Collective bargaining opens up the channel of communication between the workers and the management and increases worker participation in decision making.
- Collective bargaining plays a vital role in settling and preventing industrial disputes.

COLLECTIVE BARGAINING PROCESS

In many companies, agreements have a fixed time scale and a collective bargaining process will review the procedural agreement when negotiations take place on pay and conditions of employment.

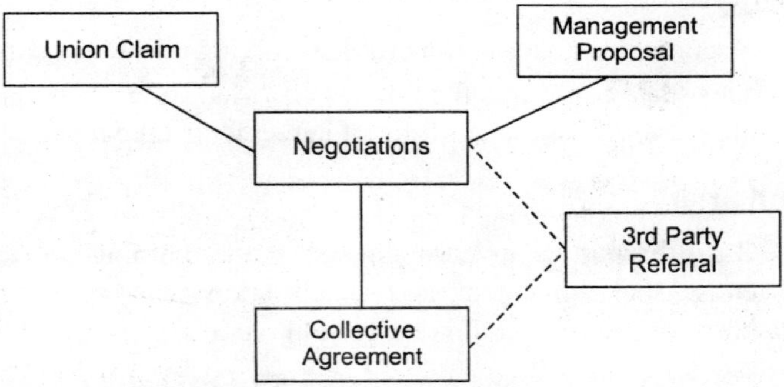

Fig. *The Process of Collective Bargaining*

- **Prepare**

  a. This phase involves composition of a negotiation team.

  b. The negotiation team should consist of representatives of both the parties with adequate knowledge and skills for negotiation.

  c. In this phase both the employer's representatives and the union examine their own situation in order to develop the issues that they believe will be most important.

  d. A correct understanding of the main issues to be covered and intimate knowledge of operations, working conditions production norms and other relevant conditions is required.

- **Discuss**

  a. The parties decide the ground rules that will guide the negotiations.

  b. An environment of mutual trust and understanding is also created so that the collective bargaining agreement would be reached.

- **Propose**

  a. It involves the initial opening statements and the possible options that exist to resolve them. In a word, this phase could be described as brainstorming. The exchange of messages takes place and opinion of both the parties is sought.

- **Bargain**

  a. Negotiations are easy if a problem solving attitude is adopted.

  b. This stage comprises the time when 'what ifs' and 'supposals' are set forth and the drafting of agreements take place.

- **Settlement**

  a. Once the parties are through with the bargaining process, a consensual agreement is reached upon wherein both the parties agree to a common decision regarding the problem or the issue.

  b. This stage is described as consisting of effective joint implementation of the agreement through shared visions, strategic planning and negotiated change.

Levels of Collective Bargaining

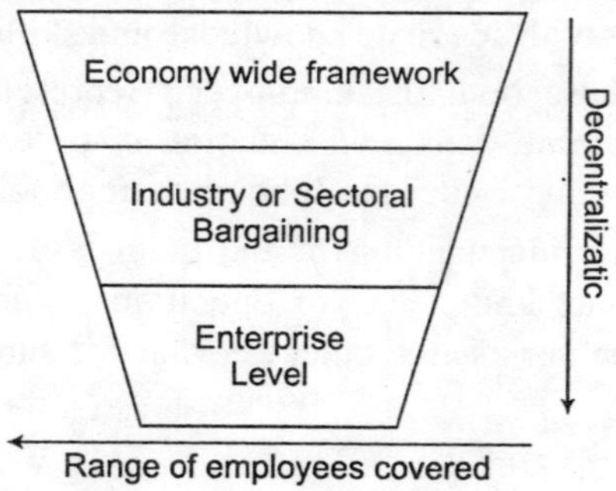

- **Economy-wide (National)**

  a. Bargaining is a bipartite or tripartite form of negotiation between union confederations, central employer associations and government agencies.

  b. It aims at providing a floor for lower-level bargaining on the terms of employment, often taking into account macroeconomic goals.

- **Sectorial bargaining**

  a. It aims at the standardization of the terms of employment in one industry, includes a range of bargaining patterns.

b. Bargaining may be either broadly or narrowly defined in terms of the industrial activities covered and may be either split up according to territorial subunits or conducted nationally.

- **Enterprise Level**

  a. It involves the company and/or establishment.

  b. It emphasizes the point that bargaining levels need not be mutually exclusive.

Problems of Collective Bargaining

- Due to the dominance of outsiders in trade unionism in the country, there is multiplicity of unions which are weak and unstable, and do not represent majority of the employees. Moreover, there are inter-union rivalries, which further hinder the process of collective bargaining between the labor and the management.
- Trade unions are having political affiliations, they continue to be dominated by politicians, who use the unions and their members to meet their political ends.
- There is a lack of definite procedure to determine which union is to be recognized to serve as a bargaining agent on behalf of the workers.
- There has been very close association between the trade unions and political parties. As a result, trade union movement has leaned towards political orientations rather than collective bargaining.

Negotiation Skill

- Any method of negotiation may be judged by three criteria

  a. Should produce wise agreement if agreement is possible.

  b. Should be efficient.

  c. Should improve or at least not damage the relationship between the parties.

- Negotiation takes place on two levels:

  a. Addresses the substance.

b. Focuses on the procedure for dealing with the substance.

- Negotiations often viewed as either:

a. "Hard"

i. Participants are adversaries
ii. Goal is victory
iii. Demand concessions as a condition of the relations
iv. Distrust others search for the single answer: the one you will accept
v. Try to win a contest of wills
vi. Apply pressure

b. "Soft"

i. Participants are friends
ii. Goal is agreement
iii. Make concessions to cultivate the relationship
iv. Be soft on the people and the problem
v. Trust others
vi. Change your position easily
vii. Make offers
viii. Search for the single answer: the one they will accept
ix. Insist on agreement
x. Try to avoid contest of will
xi. Yield to pressure

- **Dangers of "positional" negotiations:**

a. Produces unwise agreements.

i. Positions tied to ego.
ii. Negotiators locked into positions
iii. As more attention is paid to positions, less attention devoted to meeting the underlying concerns of the parties.

b. Arguing over positions is inefficient.

i. Incentives to stall settlement
ii. Agreement requires concession.

c. Endangers ongoing relationships

i. Contest of will.

ii. Anger/resentment may result from concessions required to reach agreement.

d. Multi-party negotiation complicate positional bargaining.

i. Varying positions make "common" position difficult.

ii. Changing position difficult.

PRINCIPLED NEGOTIATION METHOD

- **Separate the People from the Problem.**

a. Negotiators on both sides of the issue bring emotion, perceptions, and values to the negotiations.

b. Misunderstandings or personal perceptions of "facts" of negotiation may lead to reactions that produce counter-reactions that leads to failure of negotiation.

- **Discussions of substance entangled with emotions of issue**.

a. Statement may be intended to identify a problem but may be heard as an attack.

b. People draw inferences from comments that become "facts" about other individual's intentions and attitudes.

- **Perceptions**

a. Conflict lies in each side's perception of the problem.

b. Ability to see the situation as the other side sees it is one of the most important skills a negotiator can possess.

c. Understanding other side's position does not mean agreeing with it.

d. One way to deal with differing perceptions is to make them explicit and discuss them.

e. Look for opportunities to act inconsistently with other side's perceptions—may lead to change of perceptions.

f. Allow "face-saving"—reconciling an agreement with principle and self-image of the negotiators.

- **Emotions**

a. Recognize and understand emotions- yours and theirs.

i. Identify source of emotions

ii. Make emotions explicit and acknowledge them as legitimate.

b. Allow other side to let off steam.
   i. Listen without responding
   ii. Don't react to emotional outbursts.
c. Use symbolic gestures.
   i. Apology can defuse emotions
   ii. Even when don't acknowledge personal responsibility.

- **Communication**
  a. Three problems
     i. Negotiators may not be talking to each other
     ii. Other side may not be hearing you
     iii. Misunderstanding.
  b. Solutions
     i. Listen actively.
     ii. Acknowledge what other side is saying
     iii. Acknowledgement is not agreement.
     iv. Talk- don't debate
     v. Speak about yourself--not about them

CHAPTER 14

TRAINING: METHODS AND MODELS

A training method can be understood as an instrument or a technique that a trainer employs for delivering training content or for facilitating the learning process, in pursuance of predetermined learning objectives. In other words, training methods are the means through which communication message reaches to the trainees

Training is a planned communication process for manpower development in bringing desired changes in knowledge, skills, attitude, behavior, *etc.* in an individual in accordance with the job requirement and for better job performance in the interest of the organization. The working can be gained through literature, interaction and others communication media where as relevant skills can only be acquired through practicing and working with the professionals. Training as a process in not so simple one. It involves a series or elements and phases inter-related to each other.

- Training means to educate a person so as to be fitted qualified, proficient in doing some job.
- **According to Miller (1979),** Training is a system for changing behaviour in order to produce an increase in quantity or an improvement in quality of an individual's contribution to the goals of an organization.
- **According to Lynton and Pareek (1967),** Training is a process by which the desire, knowledge, attitude, skill and ideas are indicated, fostered and reinforced in an organism.
- **Aims of training**, Bringing desirable change in the behaviour of the trainee or the learner.
- Changes in knowledge, skill, attitude, values, beliefs and understanding.

Functions of Training Methods

- Assist in creating an appropriate learning environment.
- Enhance participation and involvement of the group.
- Promote group interaction and cohesion.

Factors Determining Selection of Training Methods

- Objective
- Subject matter
- Special characteristics of the training group
    a. Composition of the group including the sex balance, age groupings, social background
    b. Level of comprehension of the group
    c. Participants' previous experience
    d. Appropriateness of the size of the group to the method
- Time allocated for the sub-topic
- Level of competence of the trainer in effective use of the method
- Availability of facilities and resources required for the method
- Stage of the programme.

Selection of Training Methods

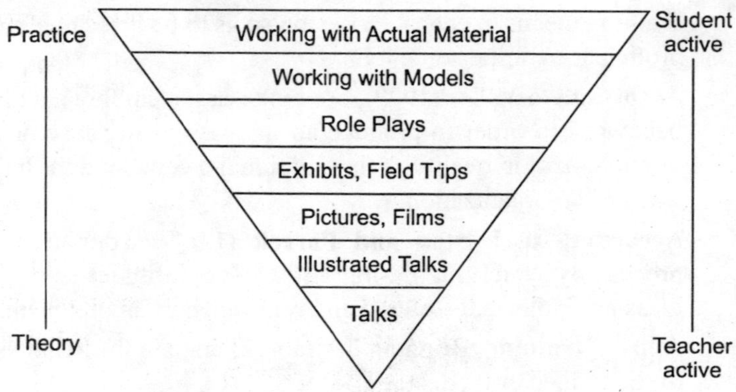

Methods of Training

- **The Lecture Method**

  a. It is one of the oldest training methods.
  b. It is most widely used.
  c. It is the most abused technique of training.
  d. It is defined as an informative discourse delivered to an audience; it clearly implies a one-way communication to a number of people by someone who is knowledgeable in his subject.
  e. This method covers most of the content under educational and training curriculum.
  f. This remains the core method even in entrepreneurship training programme.

  Advantages:

  a. Time saving
  b. Possibility of repeating the points for emphasis
  c. Economy in terms of coverage
  d. Flexible group size
  e. Safer approach
  f. Easy trainer replacement

  Disadvantages:

  a. One-way communication
  b. Repetition
  c. Lack of group participation
  d. Passive audience
  e. Boredom
  f. Attention span
  g. Lower retention rate
  h. Lack of feedback

  Attempts are made to convert the disadvantages to advantages considerably for lecture effective

  Ex. Lack of feedback is a disadvantage which may be converted to an advantage by incorporating the scope of discussion with the participants followed by lecture. Modern technology provides a wide range of aids to the speaker with the help of which the lecture method could be made more interesting, absorbing and meaningful.

The aids can be used:

- An overhead projector and slides.
- 35mm transparency projector and slides.
- Flip charts.
- Audio cassette and audio-visual.
- Video cassette recording.
- Demonstration, both behavioral and skill related.
- Preparation of the lecture itself by the trainer.
- Allowing time for questions and discussion at the end of the talk.
- Allowing interruptions for questions during the talk.
- Allowing interruptions for questions and for discussion during the talk.
- 3 components lecture
  a. Introduction
  b. Presentation of the body of lecture
  c. Conclusion

- **Conference**
  a. The literal meaning of the word conference means consultation.
  b. It is usually a highly structured device for conveying a message on a large scale often to a big gathering representing different levels of functioning amongst the participants from a similar background.
  c. The normal procedure adopted for conducting conference is to prepare the agenda notes and the working paper and invite experts to contribute papers for the conference.
  d. The plenary session which is held in beginning to spell out the objectives, purpose, the scope and specific items for discussion in the conference.

e. It provides a general forum for the participants.
f. It also helps in sharpening the focus of discussion during the course of the conference.
g. While all the delegates appear for paper presentation they are later split into smaller groups for discussion on specific items.
h. The sub-group prepares the reports which are once again discussed and a final report of the conference with suggestions for action is finalized.
i. The chairman is a key figure in the conference.
j, He is expected to have complete control over both the speaker and the audience.
k. The major task of the chairman is to sum up at the end.
l. Thus much of the success of a conference depends upon the preparatory work done for conducting it.
m. Success also depends on the selection of the chairman, delegates and expert for paper presentation.
n. The method could be effectively be used to collect ideas before embarking on a project.

- **Seminar**

a. The word is derived from the Latin for seed plot and suggests that it is a well prepared situation for sowing, nurturing and developing ideas.
b. The method combines the advantages of large and small group discussion.
c. The pattern of paper presentation is followed by discussions in a small group or in the entire group.
d. The number of participants in a seminar should be such that it can be easily divided into small syndicate groups.
e. After the initial input from a speaker, the seminar breaks into separate syndicate groups to examine the task and report back on the same topic to the full group session.
f. Handouts summarizing the main points made by the speakers are useful at the end of the seminar session to provide reference material for members.

g. Winding up a seminar involves highlighting different points on different topics.
h. In some cases discussion may take place at the end to decide the future plan.

The components of the seminar are:

a. Presentation
b. Trainer
c. Speaker
d. Participants
e. Discussion

- **Panel Discussion**

a. A panel is a small group discussing a topic in a large group situation.
b. A group of three to six persons, who have special competence in the subject and the ability to express themselves, holds a purposeful conversation under the leadership of a moderator in front of an audience.

The main purpose of a panel discussion is to

- Identify, explore and clarify issues and problems.
- Bring several points of view and wide range of informed opinion to an audience.
- Gain an understanding of the component part of topic and identify advantages and disadvantages of a course of a action.
- The method has two important aspects i.e. organization of the panel discussion and the process of the panel discussion.
- In organizing the panel emphasis is laid on the selection of the topic, the panelist and the moderator.
- The role of all the three elements is crucial for the success of the method.
- The process of panel discussion starts with an introduction of views by the panelist, discussion on issues among panelist, discussion with participants and finally summing up by the moderator.

- Theme is an important factor that should be selected for presentation through the panel discussion method.

- **Workshop**
  a. The focus of this method is product centered and on the specificity of a task to be accomplished.
  b. It is a method which aims at identifying the problems in order to find solution to them.
  c. The objective is to obtain contributions from all individuals who are affected and thus to optimize the resources available to solve the problems and plan future action.
  d. It is set up to tackle specific operational problems.
  e. The participants play a key role in making the workshop a success.
  f. It requires a highly motivated group which is aware and concerned about the problem and on which can provide volunteers for strenuous schedules of work.
  g. It demands an idealistic level of commitment and will to contribute significantly to task achievement.
  h. This is why, workshop, most of the time; continue even after dinner and late through the night.
  i. The workshop ends up with a definite plan of work for implementing solutions of a particular problem or tasks.
  j. For a workshop to be successful it requires strict homework in terms of planning the venue, selection of experts, consultants, provision of a high order of secretarial assistance and facilities like cyclostyling, typing, Xeroxing, a small library for reference etc.

- **Buzz Group**
  a. Buzz group is sub-divisions of a large group.
  b. The interaction is confined to a limited number of participants ranging from five to ten.
  c. The deliberation takes place in a small group on a given topic which is presented by the leader of the sub-groups.

d. A variation of the buzz group method is the "66" technique wherein six people are given six minutes to discuss the topic at hand.

e. The focus of this method is to provide an opportunity to his/her opinion/thinking almost everyone to express on the topic.

f. It also reduces per pressure by limiting the number of persons in a sub-group.

g. The method may be used in conferences, seminars, workshops or even for organizing group discussions on a variety of topics.

- **The Syndicate Method**

a. This method is used extensively in management programmes for executives of proven abilities.

b. It combines the role of a 'learner' and 'teacher' together to create an atmosphere of self-imposed discipline.

c. It basically consists of dividing the trainees into a number of groups to work on different aspects of the topic.

d. This small group is known as syndicate.

e. Generally each syndicate is given a brief and carefully prepared background paper by the instructor.

f. The syndicate discusses the issues involved in the subject area assigned to it and prepares a paper.

g. The chairman/leader of each syndicate presents his report separately, followed by questions raised by each syndicate on the recommendation of other syndicates.

h. The trainer acts mainly as a resource person in providing necessary information and facilitating their thinking on a particular topic.

i. The participants learn from their own participation and from the experiences the trainers and other members bring to syndicate discussions.

- **Brain Storming**

a. This method aims at generating new ideas on a problem.

b. This discussion technique emphasis on creative thinking in a group ideation process.

c. It is normally conducted in two phases known as green and red.
d. The green phase allows the participants to be non-restrictive in projecting their ideas without evaluating their ideas.
e. In the later phase, each idea may be scrutinized in terms of topic, utility, relevance, realness, and possibilities, etc.

The four basic rules followed in this method are:

a. Judicial judgment is ruled out.
b. Freewheeling's welcome.
c. Quality is wanted.
d. Combination improvement sought.

Key elements in brain storming:

a. Trainer
b. Participants
c. Recorder

- **Case Method**

a. This is an open house method.
b. The participants are confronted with a human situation documented in the form of a case and reflect their viewpoints on the situation.
c. This is used to develop the participants' skill in observation of events and judicial selection of alternatives.
d. It also helps in developing an effective way of thinking about events and systematic application of knowledge.
e. This method does not lend itself to the teaching of established theories or principles of working; rather it provides scope to develop a set of principles out of the responses that the trainer receives during analysis and discussion.
f. The effectiveness of case method depends upon the preparation of the case and leading the case in a classroom situation by the trainer.
g. The trainer is expected to perform the role of facilitator rather than a teacher.

g. The case is dealt with in a variety of ways.

h. Sometimes the case may consist of the lead questions at the end.

- **Role Play**

a. Role play is a method of human interaction that involves realistic behavior in imaginary situations.

b. It essentially consists of an interview or a series of interviews where-in the participants are provided with a role-brief and a particular set of circumstances to be enacted.

c. The way the role-brief is carried out depends upon one's role and how one believes in handling particular situation.

d. The role player develops a strategy on how to approach the task.

e. One of the common aspects identified in this method is the idea that role playing involves action, doing and practice.

f. The technique makes it possible for individuals and groups to improve their effectiveness not by talking about a problem but by actually doing something about it.

g. Role play has wide utility in leadership and management development, communication skills, improvement of inter-personal relationships, team building etc.

h. It can provide an interesting medium for observing different behavior patterns and evaluating their experiencing effects.

i. The basic training theory in role-playing including ingredients like:
   - Learning by doing.
   - Learning through imitation.
   - Learning through observation and feedback.
   - Learning through analysis and conceptualization.

j. There are at least four common dimensions of role-play structure:
   - Extent of active involvement

- Degree of situational structure
- Amount of role multiplicity
- Degree of non-verbal emphasis.

- **Simulation Method**

a. Training through structured experiences/exercise is increasingly becoming a popular method of training and is used widely for a variety of management training programmes.

b. Learning through one's own action, behavior and experiences, is the core aspect of the method.

c. In this method, the trainees are actively engaged in performing a goal oriented task and learn both from performance and from evaluating the performance process.

d. The trainees performance is based on a definite task assigned to them in terms of structured exercise.

e. These exercises are developed scientifically and are already tested for validity and reliability.

f. The objective of structured exercises could be either developing skills in relation to the affective domain.

g. Experience-based learning begins with experiencing by way of involvement in some activities and ends with finding the relevance of experiences generated.

h. It follows an inductive process wherein a person engages in some activity, looks back at activity critically and analyses his back home situation.

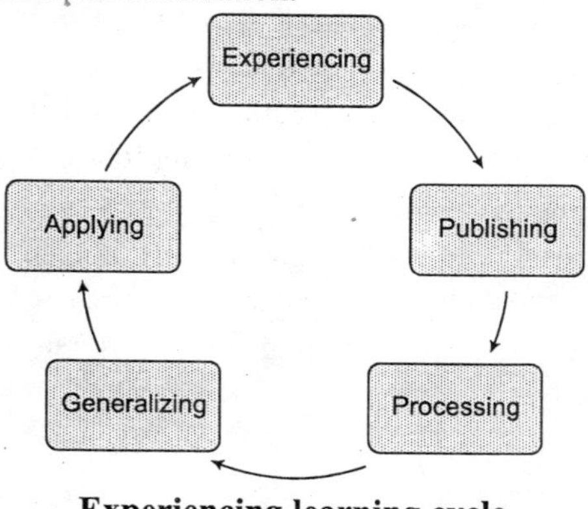

Experiencing learning cycle

Stages of Experiencing Learning Cycle

- **Experiencing**
    a. The process begins with experiencing in which the participant becomes involved in an activity or behaves in some way or he does, performs, observes, sees and says something.
    b. This initial experience forms the basis of entire learning process.
    c. In this phase the trainer is engaged in 'conducting' exercises.
    d. In any way if it stops at this point, it becomes a game, a time spent joyfully and even something of interest.
- **Publishing**
    a. The next stage is publishing their reactions and observations about the activities performed.
    b. This is the data generating stage that leads to processing.
- **Processing**
    a. In the processing stage a systematic attempt is made to examine the commonly shared experiences of the participants.
    b. The participants are helped to understand the dynamics of the activity.
- **Generalizing**
    a. In stage of 'generalizing' the participants start making an attempt to find out the possibilities of events and happening of activity and relates them to everyday life.
    b. In other words something which happens here is not just happening 'here' but happening 'there' as well.
- **Applying**
    a. Finally, the participants are helped to apply to generalization in their real work/back home situation.
    b. The actual application of learning is a new experience for them.

c. The learning through experiencing may require verification and experimentation of the learnt behavior before being accepted and acted on.

d. The experiential cycle, therefore, acts as a guideline to facilitate learning.

e. It is important to understand the significance of the learner, the structured exercises and the facilitator.

f. This method may suffer a great set-back if it is used in a mechanical way.

- **Laboratory Method**

  a. The laboratory method of training is process-oriented learning wherein the participants learn by sharing their experiences, particularly those generated in the group.

  b. The method is also termed as sensitivity training, T-group, D-group, and l-Group etc.

  c. The basic assumption in this method is efficiency and productivity of the group depends more often on the manner in which people work together than in their technology skill.

  d. The best way to understand this is to examine what is going on in the group, the 'here' and 'now' which provides rich data for learning.

  e. In this method, the participants interact in an unstructured situation, the group is not defined, leadership is not offered by the trainer, behavior is not prescribed and the group evolves through a natural process.

  f. The laboratory offers a climate conducive to open discussion, learning and encourages trusting behavior.

  g. A major goal of such a method is to contribute towards personal growth through increased self-awareness and interpersonal competence.

  h. Exploration of own behavior and experimenting with new behavior in the process and acceptance or rejection of a set behavior are a continuous phenomena in the laboratory method.

i. This is especially useful in creating an impact of the affective domain of a person.
j. Besides this method, there are a number of other methods used in different programmes: fishbowl, vestibule, transactional analysis, training within industry (TWT), field trips, assignment, job rotation, in-tray exercises, transcendental meditation (TM), coaching, programmed learning, prescribed reading, distance learning, induction training, computer based training, apprentice training, correspondence etc.

- **Business Games**
  a. It is one of the simulation methods used in top management executives training.
  b. Certain games are played by the participants.
  c. Those games are necessarily from the area of business management.
  d. Therefore, the method is titled as business games.
  e. The player in this method is the managers and other executives of high caliber.
  f. The trainers start with the analysis of instructional objectives to be achieved.
  g. The next item for the trainer is to have a thorough knowledge about the participants for whom the course is designed.
  h. The trainer is not merely an announcer but also a strict observer of the whole thing.
  i. He may also record his observations on paper.
  j. The game generates data for analysis.
  k. The business includes operations for decision-making.

MODELS OF TRAINING

Training models are basically classified into five types:

- Simple model of the training process.
- Elaborated model of the training process.
- Miller's model of the training process.

- Systems model of the training process and
- Spiral model of the training process.

Simple Model of the Training Process

- A model of training in its simplest form is presented in figure.
- It is the simplest training process where improvement is a dependent variable, participants and organizations are independent variables.

Elaborated Model of the Training Process

- A model of training in complex form is presented in figure 2.
- Training may be a temporary system such as an occasional programme or eminent institution like training department.
- The trainer in both the systems will learn through various opportunities available for checking their effectiveness.
- Hence, the independent and intervening variables also become dependent variables.

Miller's Model of the Training Process

- A model conceives the training as an interdependent and interrelated process.
- The five phases from sub-processes which are highly interrelated and interdependent on all other sub-processes as well as on the total process are shown following figure
  a. **Needs assessment process:** Many times training need assessment in HRD training is organized in a premature way results in poor performance. Hence, proper training need assessment is very essential objective.
  b. **Objective setting process:** In many training programmes objectives stated are not closely related to the real training needs, very few and immeasurable leads to poor results. Hence proper objective setting is must.
  c. **Design process:** Several training programme designed slightly related to real training needs or real life situation

and circumstances faced by the trainees. Hence, it needs proper attention.

d. **Implementation process:** Well-planned training programmes ends with poor implementation in most of the cases. Hence, proper implementation of training programme is important.

e. **Evaluation process:** This process in several occasion leads to general and less precise because of poor efforts. Hence, evaluation based on objectives is essential.

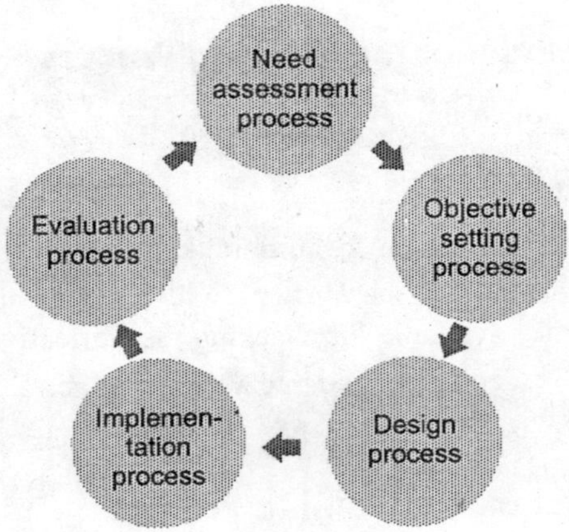

Systems Model of the Training Process

- This is a model of training extension workers.
- The process of training starts with questioning the basic assumptions which have governed our HRD approach.
- So an analysis of strength, weakness and other things analysis is called for to enable us to have new conceptualization, both of which must be based on farmers needs.
- Training of extension personnel for HRD during the next millennium must take into account the needs of broad based agricultural extension system to induce greater professional competence to serve farmer better.

- The model describes seven series of interdependent components functionally linked and incorporated into the total
- **Service action:** Any training must prepare extension personnel for service action in the field. If it does not prepare, it needs immediate reorientation for action
- **Strategic budget:** Many times training strategies overlook the budget provisions. Cost effective training strategies are essential. Hence, use locally available resources in effective ways.
- **Solved content:** The curriculum designed should help to solve the problems of clients. Hence, it requires deliberate efforts to develop a training programme with solved content.
- **Securing diligence:** Hard work with persistent effort is required for organizing extension training whether it is need assessment, objective setting, design, implementation or evaluation of training programme. No short-cut method is available for securing better results.
- **Superior extension:** Either trainee or a trainer should keep in mind that the aim of training intervention is to make extension personnel perform better, irrespective of the nature of training.
- **Serve farmers:** The farmer's requirements have to get outstanding over extension personnel requirements while training. Outstanding service should be farmer first.
- **Sustained goal:** The inner drive, impulse and intention of the person to do something in a certain way should be developed in sustained form to satisfy the needs.

Spiral Model of the Training Process

- The model helps people on the job in an organization.
- Spiral shows the phases participants pass as they learn and also how they return to enhanced capabilities.
- In this process inputs are shown as arrows in the model originating outside the spiral are inputs of the training institutional arrows originating inside the spiral, depict inputs of the work.

- Again the arrows are merely visual conveniences.
- However the arrow itself comprises continuous spiral feedback systems.
- The model comprises a training programme as an entity, as each of the series of events that make up the programme.
- We also use this as an explicate in the training process, further focusing in turn on participants organizations and finally on training systems.

BIBLIOGRAPHY

Alderson, J. C., Krahnke, K. J. & Standfield, C. W. (Eds.). (1987). Reviews of English language proficiency tests. Washington, DC: TESOL

Allan, D. (1992). Oxford Placement Test. Oxford: Oxford University Press

Allport, G. W. (1990) The Composition of Political Attitudes. Amer. J. Sociol., 35, 220-238.

Andrew, F.M. (Ed.) (1979) Scientific Productivity, and Cambridge: Cambridge University Press & UNESCO.

Anup Katoch; Fundamentals of Rural Sociology and Educational Psychology. Reading material.

Balasubramanian, T. (1989) A Textbook of Phonetics for Indian Students, Orient Longman, New Delhi.

Banerjee, M. (1984) Organization Behaviour. Allied Publishers Pvt. Ltd., Bombay.

Barnes, R. (1992). Successful Study for Degrees. London: Routledge.

Berlo, D. K. (1960) An Introduction to Theory And Practice- The Goal of Interpersonal Communication, Hoolt Rinhart And Winston Inc., New York

Bernard, L. L. (1930) Attitudes, Social. In Encyclopedia of the social sciences, Vol. 2, ed. by E. R. A. Seligman and A. Johnson. New York: Macmillan. Pp. 305-306.

Biber, D. (2006). University Language: A Corpus-based Study of Spoken and Written registers. Amsterdam: John Benjamins. (1999). Longman Grammar of Spoken and Written English. Harlow: Longman.

Bogardus, E. S. (1925). Social Distance and its Origins. J. Appl. Sociol., 9, 216-226.

Chave, E. J. (1928) A New Type Scale for Measuring Attitudes. Relig. Educ., 23, 364-369.

Chitamber, J. B. (1990) Introduction to Rural Sociology. Wiley Eastern Pvt. Ltd, New Delhi.

Chitamber, J. B. (1990) Introductory Rural Socialogy, Wiley Eastern Ltd, Delhi.

Collins COBUILD (1996). Grammar patterns 1: Verbs. London: HarperCollins.

Comfort, J. (1995). Effective Presentations. Oxford: Oxford University Press.

Dahama, O. P. and Bhatnagar, O. P. (1987) Education and Communication for Development, Oxford and IBH Publishing Co. Pvt. Ltd, New Delhi.

Dayal, I., and Thomas, J.M. (1968) "Operation K.P.E.: Developing a New Organization," The Journal of Applied Behavioral Science, 4(4),473-506.

Dey, P. K. (1997) Perspective in Mass Communication, Kalyani Publisher, Ludhiana.

Dictionary.com Unabridged Based on the Random House Dictionary, © Random House, Inc.2011.

Don Hellriegel, John W.Slocum, Jr., and Richard W.Woodman, Organizational Behaviour, 5th ed., West, St.Paul, Minn, 1989.

Dr Moya O'Brien , *Improving Your Child's Self Esteem.*

Drew, S. & Bingham, R. (1997). The Student Skills Guide. Aldershot: Gower.

Droba, D. D. (1933). The Nature of Attitude. J. Soc. Psychol., 4, 444-463.

Dyer,W.F. (1977). Team Building Issues and Alternatives. Massachusetts: Addison.

Educational Testing Service (1992). TOEFL Test of Written English Guide Princeton, NJ: Educational Testing Service.

Epton, S.R., Payne, R.L., and Pearson, A.W. (Ed) (1983) Managing Interdisciplinary Research. Chinchester: John Wiley & Sons.

Festinger, L. (1957). *A Theory of Cognitive Dissonance.* Stanford, CA: Stanford University Press.

French, W.L. and Bell, C.H. (1983) Organization Development. New Delhi: Prentice –Hall. Hindustan Times 17 October 2002 Lynda Gratton, and Tamara J. Erickson 2007. Eight Ways to Build.

Goodale, M. (1998). Professional Presentations. Cambridge: Cambridge University Press.

Guffey, Mary Ellen (2000) Essentials of Business Writing. Ohio: South-Western College Pubg.,

Halliday, M. A. K. (1989). Spoken and Written Language (2nd ed.). Oxford: Oxford University Press.

Harmer, J. & Arnold, J. (1978). Advanced Speaking Skills. Harlow: Longman.

Hasson, Gill. (2012) Brilliant Communication Skills. Great Britain: Pearson Education.

Hersey P. and K. Blanchand (1982) Management of Organizational Behaviour: Utilizing Human Resources. Prentice - Hall of India Pvt. Ltd., New Delhi.

Hornby, A. S. (2000). Oxford Advanced Learner's Dictionary of Current English (6th ed.). Oxford: Oxford University Press. Huckin, T. N. & Olsen, L. A. (1991). Technical writing and Professional Communication for Non-native Speakers of English. New York: McGraw Hill.

http://www.attitudeschool.com/effective-public-speaking-voice modulation/

http://www.lifehack.org/articles/lifestyle/63-ways-to-build-self-confidence.html

https://en.wikipedia.org/wiki/Group_dynamics

https://en.wikipedia.org/wiki/Time_management

Jeffrey Pickens (2005) Attitudes and Perceptions, 47688_CH03_043_076.qxd.

K.C. Ravi, Personality, Leader and Leadership Teaching Notes On Course No. AEX 101

Katz, D., and Allport, F. H. (1931) Students' attitudes. Syracuse, N. Y.: Craftsman Press. Pp. XXVUI + 408.

Kayfetz, J. L. & Smith, M. E. F. (1992). *Speaking Effectively: Strategies for Academic Interaction.* Boston, MA: Heinle & Heinle.

Khan, P. M. (2005) Terminology of Extension Education, Agrotech Publishing Academy, Udaipur (Rajasthan).

Kratz, Abby Robinson (1995). Effective Listening Skills. Toronto: ON: Irwin Professional Publishing,

Leadership Style: Human Resource Management, 105 FoCARS (Foundation Course for Agricultural Services) NAARM Digital Repository Course Material, Hyderabad, Telangana.

Leagans, J. P. (1958) Developing Professional Leadership in Extension Education, Cornel University, Ithaca.

Likert, R. (1933). Technique for the Measurement of Attitudes. Arch. Psychol., 22 (140), 55.

Lrcture notes on Communication Skills For Undergraduates, Amoud University, Amoud, Borama, Awdal, Somalia.

Luthans F, Organisational Behaviour, IRWIN/McGRAW-HILL 1998.

Lynton, P. Rolf and Pareek Udai (1967) Training for Development, Richard D, Irvim, Inc. and Dorsey Press, Home Wood, Illinosis.

Mamatha Bhatnagar and Niti Bhatnagar (2011) Effective Communication and Soft Skills. Person Education.

Minakshi Prasad Mishra, Pranita Tripathy and Ranjita Patra, Lecture notes : English Communication Skills (Theory).

Mohan Krishna and Meera Banarjee (1990) Developing a Communication Skills, Macmillan India Ltd, New Delhi.

Monippally, Matthukutty, M. (2001) Business Communication Strategies. New Delhi: Tata McGraw-Hill Publishing Company Ltd.,

Narayanaswamy V. R. (1979) Strenthen Your Writing, Orient Longman, New Delhi.

Notes on Interpersonal Communication: Course no. AEM-102: Communication of Agricultural Innovations, Post Graduate Diploma in Agricultural Extension Management (PGDAEM): National Institute of Agricultural Extension Management, Rajendranagar, Hyderabad – 500 030, Andhra Pradesh, India

O'Connor, J. D. (1980). *Better English Pronunciation* (new ed.). Cambridge: Cambridge University Press.

Omollo- Odhiambo, Communication Skills- Ccs 001, University Of Nairobi, Ppt

Pareek, U. (1966) Studies in Rural Leadership, Behavioral Science Centre, Delhi.

Prasad, H. M. (2001) How to Prepare for Group Discussion and Interview. New Delhi: Tata McGraw-Hill Publishing Company Limited.

R.V.S. Rao Group Dynamic and Team Building: Human Resource Management, 105 FoCARS (Foundation Course for Agricultural Services) NAARM Digital Repository Course Material, Hyderabad, Telangana.

Raman, Meenakshi & Sangeeta Sharma (2011). Technical Communication: Principles and Practice. Second Edition. New Delhi: Oxford University Press.

Ray, G. L. (2004) Extension Communication and Management, Naya Prakash, Calcutta.

Reddy, A. A. (1976) Extension Education, Shree Lakshmi Press, Baptala.

Rignall, M. & Furneaux, C. (1997). *Speaking.* London: Longman.

Rural Sociology and Educational Psychology with Relation to Extension Education. Department of Extension Education, TNAU, Coimbatore.

Rutherford, Andrea J. (2007) Basic Communication Skills for Technology: Second Edition. Delhi: Pearson Education.

R.V.S. Rao, Time Management, Human Resource Management, 105 FoCARS (Foundation Course for Agricultural Services) NAARM Digital Repository Course Material, Hyderabad, Telangana.

Sandhu, A. S. (1983) A Textbook of Agricultural Communication: Process and Method, Oxford and IBH Publishing Co. Pvt. Ltd, New Delhi.

Sharma, B.R. (1971) Occupational Values and Business: A Cross Cultural Comparison. New Delhi, Indian Academy of Social Sciences.

Sharma, B.R. (1976) Occupational Goals of Professional in the Marketing: Methodology for Cluster Analysis (Paper Presented in Faculty Seminar). Jamshedpur, Xavier Labour Relations Institutes.

Sharma, R. C. & Krishna Mohan (2007) Business Correspondence and Report Writing: Third Edition. New Delhi: Tata McGraw-Hill Publishing Company Limited.

Singh, P. (1982) Leadership Styles: Changing Indian Scenario ASCI Journal of Management, Vol.11, and No.2

Singh, P. and Dass, G.S. (1977) Organizational Culture and its Impact on Managerial Remuneration. Decision. January 1977.

Singh, P. and Dass, G.S. (1977) Organizational Culture and its Impact on Commitment to Work ASCI Journal of Management, March 1977.

Sinha, P.D. (1982) Functional Approach to Leadership. The Hindu, June 23, 1982.

Supe, S.V. (1999) An Introduction to Extension Education, Oxford and IBH Publishing Co. Pvt. Ltd, New Delhi.

Symonds, P. M. (1927). What is an Attitude? Psychol. Bull., 24, 200.

Thurstone, L. L., and Chave, E. J. (1929). The Measurement of Attitude. Chicago: Univ. Chicago Press. Pp. xii + 97.

Verma, K. C. (2013) The art of Communication, Kalpaz.

Vijayragavan J., Premlata Sing and Monika Wason (2003) Management Development: Training Modulation for Extension and Development Professionals,IARI, New Dehli.

White, D.D. and D.A. Bednar (1986) Organizational Behaviour: Understanding and Managing People at Work. Allyn & Bacon, Inc., 7 Wells Avenue, Newton Man.